WORKPLACE WELLNESS

WORKPLACE WELLNESS

The Key to Higher Productivity and
Lower Health Costs

Carol Bayley Grant
Robert E. Brisbin

JOHN WILEY & SONS, INC.
New York • Chichester • Weinheim • Brisbane • Singapore • Toronto

A NOTE TO THE READER:
This book has been electronically reproduced from digital information stored at John Wiley & Sons, Inc. We are pleased that the use of this new technology will enable us to keep works of enduring scholarly value in print as long as there is a reasonable demand for them. The content of this book is identical to previous printings.

This text is printed on acid-free paper.

Copyright © 1992 by John Wiley & Sons, Inc.

All rights reserved. Published simultaneously in Canada.

No part of this publication may be reproduced, stored in a retrieval system, or transmitted in any form or by any means, electronic, mechanical, photocopying, recording, scanning or otherwise, except as permitted under Sections 107 and 108 of the 1976 United States Copyright Act, without either the prior written permission of the Publisher, or authorization through payment of the appropriate per-copy fee to the Copyright Clearance Center, 222 Rosewood Drive, Danvers, MA 01923, (978) 750-8400, fax (978) 750-4744. Requests to the Publisher for permission should be addressed to the Permissions Department, John Wiley & Sons, Inc., 605 Third Avenue, New York, NY 10158-0012, (212) 850-6011, fax (212) 850-6008, E-mail: PERMREQ@WILEY.COM.

Printed in the United States of America

Library of Congress Cataloging-in-Publication Data

Grant, Carol Bayly.
    Workplace wellness : the key to higher productivity and lower health costs / Carol Bayly Grant, Robert E. Brisbin.
      p. cm.
    Includes bibliographical references (p.   ) and index.
    ISBN 0-471-28422-X
    1. Health promotion. 2. Occupational diseases—Prevention.
3. Industrial hygiene. I. Brisbin, Robert E., 1946-
II. Title.
RC969.H43G75   1992
613.6'2—dc20   92-6056
                                                          CIP

10 9 8 7 6 5 4 3 2

Dedicated to:
Zella Ibanez. My first corporate client.
An inspirational forward thinker.

Contents

PREFACE / ix
INTRODUCTION The Physician's Perspective
Elizabeth B. Lanier, M.D. / xi

CHAPTER 1 The Cost of Illness / 1
CHAPTER 2 The Wellness Director—Qualifications and Duties / 6
CHAPTER 3 Developing a Wellness Policy / 17
CHAPTER 4 The Wellness Budget / 25
CHAPTER 5 Developing a Fitness Program / 31
CHAPTER 6 Smoking Cessation and Substance Abuse Programs / 48
CHAPTER 7 Weight Management and Nutrition Education / 55
CHAPTER 8 Stress Management / 62
CHAPTER 9 Prenatal Wellness, Childcare, and Elder Care / 67
CHAPTER 10 Hiring, Physical Examination, and Claims Management / 73
CHAPTER 11 Insurance Options / 82

CHAPTER 12		Ergonomics / 92
CHAPTER 13		Employee Incentive and Recognition / 106
APPENDIX 1		Survey Questions / 110
APPENDIX 2		Methods of Tracking Costs and Costs Savings / 125
APPENDIX 3		Fitness Training / 132

RESOURCES / 142

REFERENCES / 148

INDEX / 149

Preface

This book is designed to provide a guide to establishing an effective workplace wellness program. The purpose of the book is to help employers reduce the cost of health and workers' compensation benefits utilization, lower absenteeism rates, and improve worker morale and productivity.

All the material in this book is based on successful methods that have been proven under actual working conditions. The information can be put into practice to reach the individual employer's desired goals. The emphasis is on presenting the principles and procedures necessary to address the most common workplace wellness problems.

An appendix addresses basic fitness and nutrition questions as well as providing tools for monitoring the effectiveness of an employer's workplace wellness program.

There is a resource section with sources of information on associations, consultants, and wellness professionals including names, addresses, and telephone numbers when available.

Introduction:
The Physician's Perspective

Elizabeth B Lanier, M.D.

Historically, America's workplaces have evolved from self-employed individuals and laborers to small busines cooperatives and finally into the modern conglomerate corporations of our industrialized nation.

Traditionally American worksite medical care has been reactive rather than proactive. Wellness programs and preventive health care plans have been figments of the imagination rather than organized components of a productive workplace.

The common scenario of workplace medical care follows a predictable sequence of events:

- A worker is injured on the job site.
- Medical care is obtained either from a nearby walk-in clinic, or the usual primary care physician of the injured worker.
- The worker is returned to the workplace after treatment and rehabilitation.
- No wellness program is instituted.
- No coordination takes place between loss control, and wellness professionals.
- In time, the worker either is reinjured or begins to exhibit symptoms of a medical condition unrelated to work duties. These symptoms inhibit the performance of duties, and/or lead to a work related injury due to a loss of concentration caused in part, or in whole, by the medical symptoms.

The disadvantages of a reactive system as outlined above are numerous. The costs of repeated medical treatment include: lost employee productivity, increased insurance premiums, reduced workforce efficiency, increased temporary employee costs, and increased training and rehire costs if the injured worker does not return to work.

When there is a lack of medical program coordination it is difficult to identify patterns of cause which can be corrected through lifestyle changes and/or workplace engineering.

In a proactive workplace, injuries and illness are prevented through a coordinated effort of loss control, with the assistance of wellness professionals. Such a system ideally begins with the initial planning of the work flow and its location.

After workplace engineering concerns have been addressed a comprehensive wellness and loss control program is developed. Planning focuses on the physical parameters of the workplace, employee duties, and non-work related activities. Perhaps most important, employee lifestyle risk factors are addressed to reduce the proclivity for injury and illness.

It is well documented in medical literature that individuals who are physically healthy and who have low levels of work related stress miss fewer days of work due to injury and illness.

Workplace environmental factors are to some extent regulated by government agencies. However, lifestyle factors have been neglected in deference to individual freedom. These factors need to be addressed through the educational portion of a workplace wellness program.

The expense and success rate of medical risk management varies considerably and, therefore, affects the choice of which factors should be included in a wellness program.

Modifiable risk factors medically relevant to wellness programs are:

- Smoking
- Lack of physical exercise
- Diet
- Alcohol and drug abuse
- Stress reduction

Smoking is usually placed at the top of the lifestyle risk list as it adversely affects the health of both smokers and their nonsmoking co-workers. In addition to the risk of lung cancer, emphysema, and heart disease, smoking also increases the risk of hypertension, ulcers, chronic bronchitis, bladder and cervical cancer, stroke, and osteoporosis. Additionally, injuries can occur as a direct result of distraction from task performance. Even when smoking is banned at the workstation, the desire to smoke can become a distraction in itself.

The reduction of smoking is easily incorporated into a workplace wellness program through education and behavior modification via group support. Medication to assist in withdrawal from nicotine can also be prescribed by the workplace wellness consulting physician.

Exercise is easily incorporated into a wellness program. Only 15% of Americans achieve the U.S. Preventive Health Services Task Force recommendation of 20 to 60 minutes of aerobic exercise three to five times per week, with strength training at least twice weekly. 40% of the population exercises minimally, and a dismal 45% percent report no regular exercise at all.

The health benefits of regular exercise are many: lowered blood pressure, stable blood sugar, stable body weight, lower cholesterol, and lower perceived stress levels.

Individuals who exercise regularly also miss fewer work days: 18% fewer for moderate exercisers, and 32% fewer for strenuous exercisers. Even modest levels of regular physical exercise add an average of two and one half years to the average life expectancy of an individual.

Exercise also increases worker productivity and concentration especially in sedentary occupations. And contrary to popular belief, age is not a barrier to the institution of an exercise program. A recent study of nursing home residents 90 to 96 years of age participating in a modest exercise program revealed at the end of four weeks a 174% increase in strength and a 48% percent increase in walking speed.

Physician evaluation may be required prior to beginning an exercise program for workers with preexisting heart disease, hypertension, diabetes, pulmonary function limitations, morbid obesity, or other limiting factors.

Diet is a popular health concern and is also easily integrated into a wellness program. The average American diet contains excessive fat, especially cholesterol and saturated fatty esters. Simple sugars, sodium, and food additives are also frequently high. Fiber and water soluble vitamins are often lacking.

High fat diets have been at least culturally associated with an increased risk of breast and colon cancer. The statistical association of elevated serum cholesterol with increased risk of coronary artery disease is well known. Diets with excessive total calories also lead to obesity, which in turn is a risk factor for the development of hypertension, diabetes, and arthritis.

Alcohol and drug abuse not only adversely affect the overall health of individual workers, these substances are also a risk factor for occupational injuries. These accidents sometimes involve innocent co-workers or the general public. Increased workplace substance abuse testing has somewhat reduced the use of controlled substances in the workplace, but wellness programs can still contribute much in this area of lifestyle change.

The myriad individual health risks of substance abuse range from hypertension and liver disease to cardiac arrhythmias and death. Substance abuse contributes greatly to the deterioration of workplace morale and social cohesion.

Stress modification occupies a final niche in the workplace wellness program, as it bridges the realms of physical and mental health. Increased stress is believed to play a role in compromising the immune system's defense against both infectious and degenerative diseases. Hypertension headaches, back strain, neck strain, asthma, diabetes, depression, and of course primary anxiety disorders are exacerbated by stress. Concentration for tasks and problem solving suffers under stressful situations, which may lead to an increased risk of worksite accidents.

Although small increments of positively viewed situational stress can be tolerated by the average worker, chronic negatively impacting stress is a risk factor which needs to be addressed by the wellness program.

The traditional role of physicians in the workplace has been episodic and specific to injury or illness. Workplace wellness programs will naturally suggest a preventive and educational role for physicians. On- and off-site screening of individuals for employment and participation status in exercise programs will also involve physicians with increasing frequency.

Occupational and industrial medicine clinics are becoming more numerous, particularly in urban areas where large employers seek streamlined medical care for sick or injured workers. Physicians employed in such settings assume key positions in workplace wellness programs as monitors of the effectiveness of prevention measures. Such clinics may also participate in substance abuse testing and cholesterol and blood pressure monitoring, and may even provide rehabilitation exercise programs in certain settings.

In the traditional private medical office or multispecialty clinic, physicians may also play an integral role in wellness programs as prehiring examiners of employee candidates and as evaluators for fitness programs, in addition to providing the usual but hopefully less frequent episodic care for illness and injury. The office based medical practitioner may assume a key role in wellness programs in smaller businesses and in rural and semi-rural communities lacking a population base to support an industrial medicine clinic. The physician may work in cooperation with the wellness program coordinator to formulate a medical risk factor management plan appropriate to the needs, strengths, and risks inherent in a particular business. Such a small, tailored program, where personal needs are met more fully, may over all be much more beneficial to individual employees than the more comprehensive but somewhat less flexible programs of large corporations.

Finally, the importance of the role of wellness director in the physician's own daily practice cannot be overstated. Here lies an opportunity to serve as a model for the community, while simultaneously reaping the benefits of the wellness program with happier, healthier employees losing fewer days of work. Physicians who address the issues of smoking, exercise, diet, alcohol, substance abuse, and stress reduction in their own offices are respected by their colleagues and patients for putting into practice the advice which they render hundreds of times weekly to others. In some medical communities, smoking physicians are decidedly the object of referral discrimination. Obesity, high stress, and excessive alcohol consumption can shorten the productive years of physicians, nurses, and other health care extenders just as it can the years of nonmedical personnel.

Breaking the tradition of a reactive system of occupational medical care and replacing it with a well designed and maintained integrated wellness program is the mission of the modern physician.

Chapter 1

The Cost of Illness

According to statistics developed by the United States Department of Labor, the nation's economy has been steadily slowing since 1969. During the decade 1960 to 1970 the net percentage increase in output per hour was approximately 2.4%. During the decade 1970 to 1980 the increase declined to 1.1%. This decline continued through the "economic recovery" years of 1980 to 1989, such that the change in output per hour for the first quarter of 1989 was -1%.

In other words, in 1989 the United States was producing goods and services less efficiently than during the height of the recession years of the 1970s, in spite of an acceptable increase in gross national product.

The advent of high technology has created industries which require workers who have excellent communications ability, well developed technical skills, and efficient managerial knowledge. But we as a nation have failed to develop an educational system capable of supplying our technical economy with applicants capable of meeting today's industry requirements. Consequently the existing skilled, albeit aging, workforce must be maintained at a peak performance level to assure maximum productivity.

Unfortunately health care costs, which are in part a reflection of the health care system's utilization and the physical health of the workforce in general, have risen dramatically.

In 1970, as a nation, we spent approximately $75 billion for medical services. In 1988 approximately $550 billion dollars, or 10% of GNP, were spent for medical services.

In order to reverse this negative trend we must reduce the unnecessary utilization of medical services, whether under the workers' compensation insurance system or under employee health insurance. And we must educate both current and future employees in the benefits of self-health management.

Many of the factors which have contributed to the increase in medical care costs are beyond the control of employers. These include: malpractice litigation, duplication of medical services, misuse of laboratory testing and technology, unpreventable health issues, advancing medical technology, and general inflationary pressures.

However, many factors which affect both the cost of medical coverage, and the efficiency of the workforce may be addressed by an employer through an effective workplace wellness program in conjunction with an effective work related injury prevention program.

Areas which can be addressed include:

Reduction or elimination of work-related traumatic injuries
Reduction or elimination of work-related repetitive motion injuries
Modified work programs for rehabilitation following serious illness or injury.
Alteration of unhealthy lifestyles through education and support
Stress management and time management
Weight management
Smoking cessation
Physical fitness, nutrition, and health education
Workstation assessment and ergonomic design
Training in safe work operations and procedures
Training in back care for maximum work life
Development of a variety of incentive and employee recognition programs to promote improved morale of the worker, which in turn promotes improved productivity

The informed employer will do everything within his or her power to maintain a consistent and healthy workforce. Without such an effort the individual businesses which make up the economy will continue to experience a net decline in efficiency, eventually falling irretrievably behind foreign competition.

In addition to aging, other considerations are dramatically affecting the composition of today's workforce. The composition of the family unit in terms of assigned responsibility has changed significantly. There are many more two-income families, more single parents with dependent children and elders, and greater ethnic and minority representation. Both employee benefit programs and workplace management systems must adapt to this diverse composition and the attendant needs.

Both health care professionals and the public in general are coming to realize that health education, coupled with lifestyle modification, is far more effective in the prevention of injury and the maintenance of good health than is expensive medical intervention. This realization is beginning to transfer to and impact the work environment.

Traditionally, employer injury and illness prevention programs have concentrated exclusively on those factors which relate directly to the duties being performed by the worker. However, many health problems which begin outside the workplace may also affect the ability of workers to perform their duties safely, or may be aggravated by the work being performed. Even when health problems do not enter into the workers' compensation system they can adversely affect the employer's profitability through increased premium costs for employee health benefits.

As a result employers are asking themselves a very basic question: How can a healthy workforce be developed and maintained?

Before management can begin to address this question, they must first take an honest, unbiased look at their workforce and determine how many of the following symptoms apply:

Absenteeism as a percentage of total employee hours over a 5 year period. How high is the absentee rate? Is it increasing?
Sick leave as a percentage of employee hours over a 5 year period. Is it increasing?
Temporary and permanent disability occurrences over a 5 year period. Increasing?
Workers' compensation costs. Experience modifiers increasing?
Employee turnover. High? Low? Average?
Labor relations. Good? Is there a strong movement to unionize. If unionized, are strikes occurring? Are they strongly supported by the majority of employees?
Communications between management and staff. Good? Are suggestions to improve quality and productivity forthcoming?

Are there repeated absences to care for elders, or children?
Are employees exhibiting signs of burnout, boredom, or lack of involvement in the success of the business?

A workplace wellness program addresses much more than the physical health of individual workers. It addresses the overall health status of the business and the workforce viewed as an interrelated entity. The healthiest and most productive businesses retain employees who feel that they are a part of the overall perspective of the operation—employees who feel that their input is recognized and appreciated by management, and who take a degree of personal pride in the work that they perform.

Therefore an employee wellness program must address, in addition to physical health questions, the question of increasing employee involvement in the continuing quality and productivity improvement of the goods and/or services provided by the employer's business.

As the trends cited above clearly indicate, a healthy workforce remains the exception rather than the rule in the United States today. The characteristics of a healthy workforce are as follows:

Development of clearly defined objectives, with a commitment to quality at the time of production
Respect and admiration for individual contributions to the overall goals of the operation
Management support of innovative approaches to problem solutions
Recognition of contributions by employees before their peers
A health policy which provides health education to shift the focus of attention away from medical intervention towards the promotion of healthy lifestyles and activities
A health policy with solid commitments to budgetary support, and attendant leadership through example by management at all levels of the operation from the top down

Finally, employers must realize that abuses of the health care system by some users and providers are to some degree within the employer's ability to control. Excessive costs of employee benefits plans are most often the result of misused services.

Once management has implemented a fully integrated overall

strategy of health care cost containment through an interrelated workplace wellness program, productivity improvements may quickly reward the employer's business with an improvement in competitive market position. When this strategy extends throughout all businesses in the nation the result may once again be worldwide economic leadership, or at the very least parity at the highest levels.

Chapter 2

The Wellness Director— Qualifications and Duties

The medium to large firm will find that their workplace wellness program will be greatly enhanced by the hiring of a full-time wellness director.

The wellness director should have sufficient formal education to understand the duties required of such a position. Many universities offer degrees in wellness, wellness management, and corporate wellness. These programs are normally offered through the physical education department. Ideally, the wellness director should have a bachelors degree, or preferably a masters degree in wellness or a related field of discipline.

In addition to a pertinent formal education, the wellness director needs a strong background in business management, human behavior, physical sciences, and counseling. Such a background may be acquired through experience, formal education, or a combination of both.

The work experience of the wellness director should be broad based. Such experience may be obtained through actual job experience, or work-study programs in conjunction with formal education. Such broad-based experience should include: fitness, nutrition, stress management, and wellness program development. A balance of education and direct experience is most desirable to ensure that the wellness director will have the ability to apply learned theories to actual daily work situations. Two to three years of field experience is very desirable.

Professional affiliations are also very desirable for any Wellness Director. Examples of affiliating organizations are:

Association for Fitness in Business (AFB)
Wellness Councils of America (WELCOA)
American College of Sports Medicine (ACSM)

One characteristic of a successful wellness director which cannot be overemphasized is the ability to motivate and work harmoniously with employees at all levels. Good verbal and written communications skills are essential.

The wellness director should be a role model for all employees insofar as health and fitness attitudes are concerned. The wellness director should exemplify a lifestyle of good health and fitness. The wellness director who is not perceived by the employee population as living by the principles the employer's wellness program expounds will not retain employee respect and the employer's program will fail to attain the desired results.

The wellness director should also express a commitment to wellness throughout the community by participating in community wellness efforts when appropriate. Such participation will reinforce the commitment of both the employer and the wellness director to the employees and their families. Family support of the employee's effort to acquire a lifestyle of health and well being is very important to the long term success of the employer's program.

FINDING A WELLNESS DIRECTOR

Because the concept of workplace wellness is relatively new, there is at present a shortfall of qualified wellness professionals.

The best way to locate a qualified professional is to advertise for a wellness director through colleges and universities offering wellness degree programs. These institutions can also serve as a good resource for information about professional organizations and trade journals that provide job placement services and information.

For many employers, hiring a full time wellness director is not an option. Such an employer will need to utilize outside consultation services and choose a wellness coordinator from within the employee population. Every effort should be made to select someone with at least some knowledge of wellness discipline, as well as the personality traits requisite for the success of the program. Such inherent talent may then be augmented by additional education, formal or informal, acquired in the course of program development and implementation. The wise employer will provide the wellness coordinator with both encouragement and assistance in the acquisition of additional wellness knowledge.

CHOOSING AN OUTSIDE CONSULTANT

The hiring of a full time qualified wellness director is not an affordable option for many employers. An affordable alternative is to hire an organization that specializes in developing workplace wellness programs. A considerable amount of time and money can be saved, as startup materials will have already been developed by the consulting firm, and the resource diversity and past experience with startup programs will prove a great asset to any employer's efforts.

Further, human nature being what it is, employees tend to label their peers, and even management members, by association of skill, talent, and knowledge with specific past performance experience. If a current employee is moved into a new position, or takes on additional new responsibility, peer credibility must be established through experience. When the new position is also part of a new employer program which will require substantial employee population participation, acquisition of such peer credibility is doubly difficult.

The outside consultant is generally perceived as an expert. Credibility of the old employee in the new position is greatly enhanced by the perceived training effect of the outside consultative advice. This will shorten the period of credibility acquisition and significantly improve the odds of success of both the program and the newly appointed Wellness Director.

QUALIFICATIONS OF AN OUTSIDE CONSULTING FIRM

The most important qualification is a reference list of other employers that have successfully utilized the firm's services. These references should always be verified. The consulting firm should provide the name of a contact person, and an address and telephone number for each reference listed. References should be queried as to specific services provided, success of programs implemented, continuation of such programs, and explanations of any failures in regard to goals set and not achieved.

In evaluating the applicability of the references given, it is important to consider the nature of the referenced operation. The greatest applicability will be found between similar operations. A wellness consultant with an outstanding track record in the construction industry may not necessarily have an equally successful record with data processing operations or other non-construction related industries.

The wise employer will always ask the consultant being considered how specific goals and problems will be addressed.

A qualified consulting firm should be able to provide the employer with a detailed description of the staff they intend to use to perform specific services. The staff should consist of educated, experienced, and trained personnel whose abilities are above average when compared to any other individual in their particular area of expertise.

The employer should be aware that some consulting firms will have one qualified person running the company, but may in fact be hiring much less qualified individuals to perform the actual services. The staff's experience should be considered as well as the firm's years in business.

A qualified consulting firm should be able to provide a comprehensive outline of their services, a proposal indicating the services most appropriate to the employer's needs, timelines for accomplishing specific goals, and an estimate of cost per module of service provided.

Often a consulting firm will use subcontractors to perform such highly specialized services as physical therapy for injury rehabilitation, or a physician to perform EKG testing and interpretation. The consulting firm should have detailed information about the qualifications of these subcontractors.

The consulting firm and all subcontractors should provide the employer with proof of malpractice or liability insurance, naming both the consulting firm and the employer as insureds.

It is important that the consulting firm demonstrate an ability to go beyond the surface of wellness issues. It must show that it is actually capable of motivating employee participants to acquire skills and abilities that will enable them to make positive lifestyle changes resulting in a higher level of health and well being.

There are a few narrowly focused wellness programs available on the market today that appear, on the surface, to provide all that is needed. They usually consist of educational materials, written questionaires, and so forth. This type of information can be useful; however, it is important to recognize that performance is what truly separates a wellness program from good information.

Without measurable success, lowered health care costs, reduced absenteeism, fewer workers' compensation claims, and increased productivity will not be forthcoming.

A comprehensive program must go beyond education and awareness. Education and awareness programs typically serve only those individuals who are already motivated to live healthier lifestyles, who are, in fact, not overutilizing medical services. However, they do not address the needs of those people who are less motivated. Therefore, an effective wellness program should also appeal to those hard to reach individuals who will benefit the most from the program.

Professional affiliations are an important part of evaluating a consultant's qualifications. Industry organizations are a valuable resource for workplace wellness programs. A consulting firm that is active in these organizations, and takes advantage of the growth offered by such affiliations, will be assured of being current with the most up-to-date information in the field. This will allow the consultant to provide the employer with the highest quality of service.

A consulting firm which has no professional affiliations may, in fact, not have the capabilities that it claims.

Affiliations listed by a consultant should also be researched and verified to ensure current membership status and the validity of the affiliating organization, and to determine the services they provide to their members.

SPECIALIZED CONSULTATIVE SERVICES

Ultimately, the primary function of any wellness director is coordination and administration. The director will rarely perform all of the actual services required in a comprehensive workplace wellness program. Therefore, it will be necessary for the director to contract with an outside instructor for some of the individual services required to fulfill the employer's wellness program needs.

The director should evaluate potential single service providers in much the same way as would be required for an overall consultant. Providers should be qualified and credentialed in their field of expertise.

Fitness

It is desirable for a fitness instructor to have a degree in physical education, exercise physiology, or some other applied physical science program. A good fitness instructor should have a thorough understanding of anatomy, physiology, kinesiology, and body mechanics. Most university level physical education programs address these subjects.

It is important to note that to be an effective fitness instructor, an individual must possess excellent "people skills." Practical experience is as important as formal education, a balanced background of education and experience is ideal.

The fitness industry has a wide range of qualified and not so qualified individuals.

The public has a tendency to evaluate a fitness instructor's ability by their physical appearance. If their body is toned or muscular it is assumed that they know all they need to know about fitness. This is not necessarily the case. In fact, the body builder is at present typically not very knowledgeable in the sciences, but primarily knowledgeable in how to use equipment. There are notable exceptions to this general observation, of course.

In an attempt to upgrade the image of persons working within the fitness industry who lack the formal education necessary to place them in a professional category, several fitness organizations have devised certification programs that provide fitness instructors with an opportunity for professional growth. These certifications tend to be very activity specific. In other words, an instructor with aerobic dance certification does not necessarily have the ability to teach weight training. The reverse is also true.

Nearly every fitness activity incorporated into an employer's wellness program will require different abilities, each instructor should possess those specific abilities necessary for the subject taught.

Organizations offering fitness certification are:

American College of Sports Medicine
National Strength and Conditioning Association
International Dance Exercise Association

When considering specialty instructors, an individual with one or more certifications from the above organizations will most likely be qualified to do the job. Ideally, an instructor with a degree and the appropriate certifications would be the best choice.

Professional affiliations are also important for these instructors, as well as references from previous positions or clients. The quality of your program is dependent upon how you and your contractors are perceived by the employee population. Your instructors must possess the image you want to project.

An additional problem with informally educated instructors who rely primarily on inherent talent and genetics is that they do not have the knowledge necessary to assess the fitness capabilities of average to below average individuals. Their experience data base is heavily weighted by their own inherent talent, as well as their individual interactions with other equally or nearly equally talented individuals. Hence they are more likely to push some employee participants beyond safe and attainable goal levels, or conversely instruct at an unnecessarily conservative level. Neither approach will enhance the effectiveness of the employer's wellness efforts, and may in fact cause injury where none existed before.

Nutrition Education and Weight Management

In 1989, Americans spent $28 billion on diet related goods and services. There is a lot of misinformation doled out to the public about nutrition and diet. It is important for the wellness director to have some background in nutrition in order to select appropriate providers for weight management and nutrition programs.

A provider should be well educated in the field of nutrition. Having a degree in nutrition, or being a registered dietician usually assures such requisite knowledge. Because our physical wellbeing is heavily influenced by nutrition, many physical education professionals are specializing in sports nutrition and weight management. An individual with a masters degree in the field, or evidence of extensive nutrition training, would also be a good choice.

Experience is essential. A provider must have experience in working with people. Good counseling skills are a requisite. They must possess an ability to communicate technical information in very simple, practical terms. It is recommended that a provider have two to three years of experience working with people in worksite, or community settings.

Wellness directors are cautioned against contracting with fad diet organizations to provide these services. These programs do not provide participants with the education or skills required to maintain lifelong healthy eating habits. Participants will experience rapid, temporary weight loss on these programs, and regain the lost weight within a short period of time.

A provider should offer a program that allows participants to acquire an understanding of basic nutrition and then apply that information to daily routine. Participants need to be taught how to develop eating habits that are appropriate for them as individuals. The wellness director should question potential providers in regard to how they plan to provide such a service.

As with all providers, professional affiliations and references are very important, and should be researched.

Stress Management

Stress Management is an area of wellness that is getting a lot of attention in the work setting. In fact, stress related workers' compensation claims are escalating at an unprecedented rate. Because there are so many aspects to stress, and how individuals are affected by it, it is often difficult to determine whether or not the stress related claim actually occurred at work.

However, regardless of the source of the stress, it is necessary for individuals to learn how to manage the stressor. Contrary to popular belief, stress cannot be eliminated. In a very real sense all external input and some internal reactions are stressful. The important point is not the stress, but how the stress is managed. An individual who acquires good stress management skills will be able to cope successfully with a variety of stressors both inside and outside the workplace. An effective stress management program will provide participants with the opportunity to learn how to manage the stressors in their individual lives.

The wellness director should question providers in regard to how their program will accomplish teaching stress management. Typical services offered for stress management include:

Time management
Communications Skills
Stress mapping
Biofeedback training
Massage and other relaxation techniques

    Providers are often not skilled in more than one of these areas. It is important to recognize that a balance between these services is necessary, and that a comprehensive program must be organized and coordinated by the wellness director. Often individual providers specializing in only one area of stress management are of the opinion that their particular specialty is all that is required. However, this is seldom the case. A variety of approaches are necessary to address the individual needs of the workplace population. Further, needs will vary between various work settings and programs should be flexible to facilitate the meeting of those specific needs.

    The education and experience of providers will vary greatly. There are not as yet degree programs in stress management, so the wellness director will have to depend solely upon the experience and references of prospective providers.

Back Care

Many companies have some form of back care education included in their loss control or safety programs. Existing back care programs can be easily incorporated into a wellness program. A wellness director should work closely with safety and loss control personnel to develop the back care program, and to maximize the opportunity to prevent future injury, as well as rehabilitate existing injuries.

    If the employer's safety or loss control program does not address back care it is suggested that the wellness director research existing back care programs in place at other companies. Quite often physical therapists or even chiropractors provide preventive back care services. These providers should be licensed and able to provide a list of references. It is important to determine during reference research how effective the provider's program was found to be by a former client. Unfortunately, some unscrupulous providers use the program as a means of gaining access to a large patient population. Referenced statistics should show a low rate of injury treatment following implementation of the providers program and services.

Smoking Cessation

There are numerous smoking cessation programs on the market. Some provide training of an employer's personnel to administer the program, others may offer outside instructors. While others offer a self-help approach through support group courses.

The basic components of any cessation program are essentially the same. The differences are in the services delivered by the provider, and the purchase price. A wellness director should research all available programs, and choose the one that is the most appropriate for the employer's workplace. Be wary of program salespersons who are reluctant to provide specific details about exactly how their program works, due to trade secrets that no other program offers. There are no secrets in smoking cessation. A tremendous amount of research has been done on the subject, and is available to everyone.

As always the wellness director should determine the qualifications of providers and request references.

Drug and Alcohol Education and Rehabilitation

These programs are often initiated through employee assistance programs which may be coordinated through community, state, or federal grant programs. Wellness directors should work closely with EAP personnel.

If the employer does not have an employee assistance program established, the wellness director should seek out certified alcohol counselors for drug and alcohol related services. Certified and/or licensed personnel can assist the wellness director in developing an appropriate program.

CHOOSING THE RIGHT PROGRAMS

Wellness has become a catchall term, much like the word *natural* in food manufacturing. One can even buy "wellness" pills and tonic in retail stores or through the mail.

It is important for an employer to recognize that in order for a wellness program or service to be effective, it must be comprehensive.

There are single service organizations that represent themselves as wellness providers. Their specific service may be valuable to, or an adjunct to a true wellness program, but should not be equated to a complete program.

For example, some health clubs offer "corporate wellness programs." This usually is a membership sales gimmick which offers a lower price structure for large numbers of individuals who sign up for membership at one time. Gyms and health clubs can be a useful part of an employer's program especially in regard to weight and aerobic dance training, but certainly do not constitute a comprehensive wellness program.

Other organizations may offer services such as health risk appraisals. Again, this is only one aspect of a true wellness program.

Medical facilities may provide wellness programs. These are generally "patient getters," or at best offer increased employer/medical personnel communication when treatment is provided to the employer's workers.

Insurance companies sometimes offer wellness programs through designated medical providers. Here the company is really trying to limit its costs by contracting for medical services at predetermined rates while gaining access to medical information germane to preexisting conditions. Such information is often quite valuable for future denial of subsequently submitted claims.

The employer should always be skeptical of any program which requires an outright purchase of goods or services, especially if the program is designed to be totally implemented by the employer's staff. "Train the trainer" programs are rarely effective, and often quite costly to purchase.

CONSULTANT GUIDELINE SUMMARY

Research malpractice and liability coverage. Ask about past claims.
Contact the local Better Business Bureau for past complaints.
Research and verify professional affiliations.
Research and verify all references.
Compare costs for services rendered.
Research and verify degrees, certifications, and licenses for validity and current status.

Chapter 3
Developing a Wellness Policy

The first step to implementing a workplace wellness program is the development of a wellness policy. This policy should express the commitment of the employer's management team to employee health and wellness. The policy should also outline the expectations of the employer in regard to the employee's involvement in the wellness program.

The following is an example of a company wellness policy statement:

(Sample Wellness Policy)

(Company name) supports the idea of workplace wellness and believes that a healthy employee is a productive employee. It is the intent of this company to provide a work environment that facilitates the personal health and well being of each employee.

(Company name) has developed a wellness program to advance employees' health and well being and to improve the quality of their lives both at work and at home.

The goals of this program are:

1. To help our employees gain the knowledge, skills, and motivation needed to achieve and maintain a lifestyle that promotes personal health and wellness.
2. To develop and maintain a workforce that is conscious of health and safety practices and, therefore, less prone to illness, or injury.
3. To reduce the cost associated with lost work time due to medical treatment for illness or injury.
4. To increase employee productivity.

The wellness program provides lectures, published materials, and activities for: stress management, nutrition, physical fitness, back care, injury rehabilitation, smoking cessation, and substance abuse recovery. These programs and activities will be available to all employees. Incentives will be awarded to participants who set and achieve specific health/fitness goals. Participation is not mandatory, but is strongly encouraged.

Employee input in planning and implementation of activities is welcome. All employees are invited to contribute ideas to the Employee Wellness Committee and to help in the planning and coordination of all wellness activities. Employee comments in the form of program evaluation are encouraged.

(Signature of Company President)
(Date)

The wellness policy must be supported by all levels of management. Support is best shown by management setting an example for the workforce to follow, as would be the case for any other aspect of the employer's operations.

A wellness program must be considered a part of normal business operations. Wellness activities should be viewed with the same importance as any other activity in which the employer and employees would normally participate.

Historically, wellness programs have been completely voluntary. Exceptions to this general approach have normally been programs designed to maintain, and/or improve the overall fitness of employees engaged in certain types of duties which by their very nature require a high degree of physical fitness, e.g., firemen, law enforcement officers, or material handling personnel.

Where such specific requirements are not part of the employee's job description an argument has been made that an employer cannot force an individual to participate in a wellness program simply to improve his or her lifestyle. Such mandatory requirements could be construed as an infringement upon an individual's right to live life as he or she chooses, even if the chosen lifestyle is ultimately deleterious.

While this argument has validity, there are certain aspects of a wellness program which can be made mandatory for all employees; these aspects relate to data collection for program planning and evaluation. Without the full collection of data from the entire employee population, it is very difficult to develop an accurate needs assessment. Also, without a thorough baseline the data comparison of wellness differences between participating and nonparticipating employees is impossible.

Chapter 3: Developing a Wellness Policy

Even though mandatory participation by all employees in the employer's wellness program is probably not feasible under most employment situations, every effort should be made to encourage full participation. In fact as the program progresses, peer pressure, as well as an employee's natural desire to be a part of a success oriented group, will bring nearly the entire employee population into the wellness program.

Often first year participation in a wellness program will run only 20-30%, and can run as low as 10%. There are two primary reasons for low participation rates:

1. The program does not meet the interests of the employee population because it is narrowly focused.

An example would be the building of a weight training facility on site without completing an interest assessment survey which determined that the majority of the employee population was in fact interested in a weight training program. This is not as simple as it would appear, however. A simple questionnaire asking employees if they would like to see the employer build a weight training facility is not the same as a true interest assessment survey. Employees will almost always respond positively to an employer's offer to build any kind of an employee facility. An interest assessment involves both the determination of employee acceptance of the proposed facility, and a commitment from the employee population in regard to its utilization.

2. A passive and/or apparently disinterested attitude on the part of company management. Such an attitude is projected when management members themselves do not make a visible commitment to the program.

Further, as touched upon earlier, any activity advocated by management must be given priority equal to that given normal company activities. It has been found that a sharing of commitment by both the employer and the employee is most effective in developing full employee population participation.

For example, weight training sessions might begin twenty minutes before the start of the work day and continue twenty minutes into the work day. Or midday activities would begin during a lunch hour and continue for a few minutes into the afternoon work period, or begin just before the end of the work day and continue after normal working hours.

By requesting employee involvement and at the same time committing to management involvement, management is communicating to the employees that the activity is important, and that it should be taken seriously.

There are essentially three phases to a comprehensive workplace wellness program:

Phase 1. Needs and Interest Assessment.
Phase 2. Implementation of specific programs designed to address the assessed needs.
Phase 3. Ongoing evaluation of the employee population in terms of goal attainment and future need. A workplace wellness program is not a static program which once implemented never needs revision. It is a fluid activity which changes as the needs of the workforce population change. At the very least certain goals and activity intensities will have to be raised to adjust for the increased physical fitness quotient of the workplace population.

All three phases should be based upon realistic goals and objectives relative to individual workplace populations, along with estimated time lines for goal attainment. Each phase must include an ongoing evaluation of methods, objectives, results , and employee acceptance to ensure maximum program effectiveness and fluidity.

Integrating these three phases into a calendar year or the employer's fiscal year is important to ensure smooth coincidence with normal budgetary projections.

All three phases are essential, regardless of the size and/or scope of any wellness program. The complexity of each phase can be modified, but all phases should be addressed to ensure long term success.

PHASE 1. NEEDS ASSESSMENT

It is important that baseline data be gathered accurately on:

Current lifestyle habits of the employees
Employee interest in wellness

Current overall employee productivity
Current Workers' Compensations costs and claims frequency
Current Health Care insurance costs and claims frequency
Current employee absenteeism, increasing, or decreasing relative to prior three year period
Current health risk and fitness levels of the employee population

To acquire the answers to phase one baseline data needs a variety of questionnaires may be used:

1. *Employee interest surveys* solicit employee input to determine the kinds of programs in which the employees would be most likely to participate. Without this information, an employer runs the risk of offering a program in which there is little likelihood of optimal participation.

The survey also gives the employee population a chance to be a part of the initial stages of the employer's wellness program development.

2. A *health and fitness assessment* gathers baseline data on current levels of employee physical fitness and health risk factors. The information gathered on each individual employee provides the wellness professional with a means of assessing the basic issues most needed by the employee population, and also provides the data necessary to design appropriate activity programs geared to the particular workforce.

Typical components of an onsite fitness field test and health screenings are:

Health risk appraisal body composition (percentage of body fat)
Blood pressure
Heart rate
Aerobic capacity (cardiovascular condition)
Muscle strength and endurance
Flexibility
Blood work, including tests of total cholesterol, HDL cholesterol, LDL cholesterol, triglycerides, and blood glucose levels

Wellness program personnel should work closely with the employer's designated health care professionals to minimize duplication of tests and screenings and insure employee confidentiality. It should also be noted that certain individuals may require special attention because of existing medical conditions that will necessitate cooperation between medical and wellness personnel.

PHASE 2. ACTIVITY PROGRAMS

Once employee needs and interests have been determined, the next step is to develop and implement activity programs which address the established needs. These activity programs may include: fitness, nutrition, stress management, injury prevention, smoking cessation, substance abuse recovery, and/or such other programs found appropriate to the employee population's needs.

PHASE 3. EVALUATION

Evaluation of the wellness program's success in attaining predetermined goals cannot be overemphasized. The program evaluation should address a variety of items, among them cost benefits accruing to health care insurance, workers' compensation insurance, absenteeism, and productivity. Generally, insurance cost reduction will require three to five years of program continuity to acrue. However, improvements in productivity and reduced absenteeism may be discernable within the first year of program inception. Employee morale improvement and program perception may also be noted early in the evaluation phase, as will increased levels of employee health and fitness and modified health risk factors within the first program year.

The results of the evaluation phase may then be used to determine modifications to the program to ensure continued employee participation and progress toward the overall wellness goals.

GOAL SETTING

Goal setting is an essential element of any wellness program. Wellness program goals should be developed with the assistance of wellness professionals and management.

The general long term goals of an employer wellness program are:

To develop a wellness policy
To design and implement a workplace wellness program

To generate full participation in the employer's wellness program by the employee population

To improve the level of employee population health and fitness, and increase productivity

While these overall goals may take some time to attain, short term individual goals and objectives should also be developed and monitored to enable the wellness program via the employer's management team and the wellness professionals to accomplish the purpose of the overall program.

WELLNESS PROGRAM COMPONENTS

There are a variety of approaches, or components available to assure the attainment of overall Wellness program goals. These include, but are not necessarily limited to:

Physical exercise
Recreational programs
Nutrition education
Weight management programs
Smoking cessation
Second hand tobacco smoke control
Alcohol and substance abuse rehabilitation programs
Stress management, psychological and physiological
Time management
Type A versus Type B behavior, and management
Relaxation techniques
Stress coping mechanisms
Improvement of verbal communications skills
Goal setting and attainment programs
Life stage transition awareness
Assertive versus aggressive behavior modeling and education
Opportunity recognition and expansion
Self imposed goal setting
Self esteem building and maintenance
Recognition of self accomplishments
Positive thinking reinforcement
Interpersonal relationship management

Parenting skills associated with work time management
Employee counseling services
Elder care options
On premises small child care
Job promotion opportunities through education
Injury Prevention
Safety and production incentive program development
Employee job/product quality commitment

Additional programs may be identified, or developed as the wellness program progresses.

It is not suggested that any one employer would necessarily offer all the programs noted above; however, it is quite possible that over a period of time all the areas noted above, and others not noted, will at one time or another be a part of a dynamic and fluid workplace wellness program.

The employer's written wellness policy may be decided entirely by management edict, or may be developed from an employee and management joint committee approach. There are pros and cons for either method of policy development. However, where program development is concerned it is most effective to form a wellness committee consisting of management and employee representation.

Chapter 4

The Wellness Budget

Many employers will hesitate to develop a workplace wellness program because they perceive such a program as requiring a large capital outlay in order to ensure a comprehensive and effective approach.

Although a wellness program will require budgetary allocation, large capital expenditures are not necessarily a requirement.

The first step in the budgetary process is to determine which elements of a wellness program are most needed and most desired by the employer's workforce. Sometimes there are discrepancies between what is needed and what is desired.

From the beginning the employer needs to involve the employee population in the development process; the first step is an interest survey.

An interest survey is an inexpensive way of starting a workplace wellness program. From the interest survey two important pieces of information will be derived: (1) which features of a wellness program are of greatest importance to the workplace population, (2) who within the employee population can be integrated as volunteer talent for the employer's wellness efforts. Most employers will be pleasantly surprised by the wellness program possibilities that can be developed without heavy monetary expenditure by drawing upon existing talent within the employee population.

The next step of a wellness program is to provide employee health screenings. Through such screenings employee interests and desires can be brought into line with their health needs. A needs assessment is a required starting point for all employers, regardless of employee population size.

Such an assessment is necessary not only from a health status viewpoint, but also to ensure that the employer maximizes the dollars spent by concentrating the program's efforts where the greatest need and interest exists for the given employee population.

For example, if a high percentage of the employees show a need to address elevated blood pressure, then this would be a starting point for the employer's wellness program. Remember, there are no generic wellness program foundations, beyond interest and needs assessments. Each workplace population is unique.

Many employers perceive the establishment of a wellness program as involving the purchase of expensive weight and fitness training equipment, as well as the setting aside of significant space for the use of such equipment. In fact, as noted earlier, this is not necessarily the case.

Certainly those employers who can afford a dedicated fitness center and whose employee population shows a sincere interest in such a center should not hesitate to proceed in its development. However, such a large capital outlay is neither a prerequisite nor an inevitable requirement of a successful wellness program.

It is important when developing a wellness program budget that both short and long term goals be considered and integrated into the overall wellness plan. This is necessary to ensure that each phase of development will build upon a prior phase, eventually leading to a comprehensive and flexible program which completely addresses the needs of the employee population.

One effective way to ensure long term growth is to track program results in a manner which indicates the shift in the employee population health risk status which is occurring as a result of the wellness program. Tracking data in this manner not only shows the effectiveness of the program but provides information which may be used to quantify the monetary benefits gained from each phase of the program. These benefits may then be translated into savings which may be reinvested in further expansion of the wellness program.

For example, let us assume that workers' compensation claims indicate a high frequency of strain and sprain injuries among the employee population as the result of material handling exposures.

Let us further assume that all loss control engineering measures reasonably available have been implemented, such that the present material handling needs have been determined to be intrinsic to the economical pursuit of the employer's operation.

Now let us introduce a wellness module on back care combined with strength and flexibility training. The results of such training may then be correlated with the reduced frequency and severity of material handling workers' compensation claims.

By knowing the past cost of such claims and the present claims frequency reduction, a dollar savings value may be assigned to the back care wellness module. And the wise employer will then use such cost savings to offset current wellness program investment, as well as to fund future program growth.

By applying these budgetary principles to the entire wellness program a breakeven point may be determined beyond which the wellness program begins to produce a cost benefit for the employer.

Let us not forget that tangible results will also be seen through the reduction of costs associated with employee non-work related health benefits. While these benefits are a little more difficult to correlate, they are correlatable and should be factored into the overall data used to track the employer's wellness program.

In this regard it is interesting to note that some states are beginning to experiment with the concept of 24 hour employee health coverage plans. Such plans combine both workers' compensation insurance and employee health benefits insurance into one comprehensive insurance program.

It is our opinion that 24 hour insurance coverage will be an option in the future, and it may in fact become a requirement via the same set of justifications which led to the establishment of the present workers' compensation system.

It is a fact that a significant portion of the United States' gross national product is being diverted to fund the medical care provided to its population.

In order to reduce and control the escalating costs of such medical care, a comprehensive medical insurance plan must be developed. To be viable such a plan would need to address the excessive litigation which has been occurring as the result of disputes between workers and employers over the work relatedness of cumulative illness and injury. Currently millions of dollars in health care benefits are siphoned off to pay legal fees associated with responsibility disputes.

Furthermore, standard treatment costs must be established. These standards can best be established by looking within the existing health care system and applying actuarial data modified to ignore cost inflation produced by excessive litigation, as well as reallocation of unpaid medical services.

Currently physicians and health care organizations absorb the cost of unpaid medical services by passing those costs on to the individuals and organizations who do pay for provided health care services. Hence an ever increasing cost profile is created.

When a 24 hour health care plan is dissected, it becomes obvious that it divides into two parts: illnesses and injuries which are work related, and illnesses and injuries which are not work related.

An employer's workplace wellness program is a form of loss control which addresses both sides of the medical services utilization question. It is applicable under current insurance plans which separate work related and non-work related medical needs, and it will be a requirement of future comprehensive plans just as work related injury prevention programs are currently a requirement of work related injury plans.

The wise employer who establishes, tracks, reinvests in, and fully develops a workplace wellness plan will not only reduce current costs significantly, but will be in the very best position to access future health care plans in the most economical fashion. Such an employer will have a plethora of competing companies vying for future insurance business.

The employer should research available community services for integration into the wellness program. Often community services can be utilized until the employer's budget will allow the community provided service to be brought into the workplace. On the other hand, space or other factors may dictate that community services are the best option even into the future. The important point is not to overlook these services during the development phase of the workplace wellness program.

Community services should not be used simply to reduce program costs, beyond what is reasonable. A miserly approach to program development will ultimately reduce program effectiveness. Certainly, utilization of community services is better than not offering that portion of the wellness program at all. However, if the budget is available, bringing the program in house will be most effective.

It must be remembered that most community programs are underfunded and therefore rely heavily upon volunteer and, too often, inexperienced workers. The employer should keep in mind the old adage that one gets what one pays for. Do not skimp on program expertise unnecessarily.

The wellness director needs to research the options available within the community so that an overall wellness plan can be implemented in the most cost effective manner. It is important to balance community services and facilities with contractor services and in house facilities.

Studies undertaken by John C. Erfurt and Kenneth Holtyn, MS (*Worker Health Program,* copyright American College of Occupational Medicine, Ann Arbor MI 48109-2054) have shown that employee participation in workplace wellness programs is severely reduced when program costs are charged back to the participating employee. Nevertheless, funding or partial funding of portions of a wellness program by participating employees may be a necessity, and is preferable to no program at all. However, this option should be regarded as a last choice.

Employee funded modules may be viable where employee desires for a program module fall outside the needs assessment data, for example, purchasing equipment for certain recreational activities or team sports.

Another funding contribution approach is to make the wellness program available to members of the community or to other employers in the immediate geographical area for a stipulated fee. Some employers may be reluctant or unable to make a commitment to their own wellness program.

Charging a stipulated fee for spouses and children of employees is another way of supplementing program funding.

While such options are viable, they do pose increased liability risks which should be researched by the employer's risk manager. Many independent insurance agents provide risk management consultation as part of their brokerage services when reviewing an employer's insurance needs. In fact, risk management overview and exposure assessment is one of the service advantages gained by an employer who chooses to deal with a sophisticated independent insurance agent.

The wellness director needs to be creative in developing and justifying funds for allocation to the employer's workplace wellness program, making certain that the greatest value is gained for the dollar spent and that program goals are not compromised.

There is a nonquantifiable gain from a well run program, and there is an equally nonquantifiable loss from a poorly run program. Certain basics may not be compromised. The employer needs to make a top down management commitment to developing an effective, well run workplace wellness program in order to realize maximum benefits from the effort expended.

Thus far a point of diminishing returns has not been identified where workplace wellness programs are in place. This is not to say that such a point does not exist. However, the more creative the program, the more successful it will be.

Training specifically identified employees to become wellness module instructors is one creative measure which will have long lasting benefits. Such an approach encourages greater employee participation and long term employee loyalty, and generally reduces employment turnover. Such an approach also has obvious cost benefits.

The question of whether or not employees should be paid for their instruction time should be addressed at the outset of budget development. Certainly if instruction will take place during normal working hours payment will be required by labor law. However, whether or not employees should be paid for non-work hour time is a question which must be individually decided by each employer. It is our opinion that if employee instructors are given the responsibility to maintain logs and records as well as creatively lead program modules, then some reasonable monetary remuneration or comparable time off should be provided. The monies invested will be well spent and certainly less costly than utilization of an outside contractor for leadership in all of the wellness modules.

The wellness director should always look to the employee population for ideas and growth possibilities regarding the wellness program. To remain effective a wellness program must have variety and flexibility. A fixed, inflexible program with little tolerance for variety will rapidly lose the participants' interest.

Full participation must always be a primary goal of any wellness program; without full participation maximum benefits cannot be realized and the per participant cost will ultimately go up to unacceptable levels.

Chapter 5
Developing a Fitness Program

Regardless of the size of the employee population it is possible to provide a variety of fitness activities. These activities should be need and interest specific.

The employee population assessments completed during phase 1 of the workplace wellness program will provide both the participants and the wellness professional with baseline information about each participant's current level of health and fitness.

Participants should be placed into activity programs that will enable them to accomplish their individual health and fitness goals.

It is important that the fitness program recommended to each participant should address the participant's individual needs in order to avoid the aggravation of preexisting medical conditions or the development of an adverse condition as the result of improper levels of training.

All fitness activities need to be performed at an intensity level consistent with the participant's level of ability and potential for goal acquisition. Therefore, individualization of the program is essential.

It is important for any fitness program to teach participants specific and individualized fitness skills. This approach will enable each participant to become more personally responsible for their own day-to-day health and well-being.

Just as a safety conscious attitude is the primary factor in reducing and eliminating work related injuries, a health and fitness attitude on the part of each participant is essential to the success of any workplace wellness program. Ultimately, each participant should take full responsibility for the maintenance of his or her own health and fitness, rather than relying or becoming dependent on a fitness instructor or personal trainer for accomplishing each exercise activity or for reaching specific goals.

This is not to say that monitoring is to be eliminated, but rather that each participant must realize his or her own very significant health responsibility.

It is important that participants engage in an activity which they perceive to be enjoyable. This is one of the reasons for taking an interest survey during phase 1 of the development of the wellness program. Often participants will get involved in programs they think they should participate in, rather than those which are in fact enjoyable for them.

For many individuals organized fitness programs or routine workout schedules are not associated with pleasure. This is the primary reason that so many people are unsuccessful in maintaining regular fitness activity. Therefore, it is important to introduce these individuals to activities which are both pleasurable and at intensity levels appropriate to their own levels of fitness.

It is true that ultimately good health and fitness are motivators in their own right. But enjoyable participation is important to ensure that those individuals who have not experienced the benefits of routine fitness exercise do not fall into the trap of participating in a program whose intensity is too great and whose activity is at best uninteresting and at worst disagreeable. Inappropriate intensity levels and uninteresting activities lead to burnout and ultimately program failure for that participant.

It is also important to recognize the interrelatedness of fitness and exercise with other components of a workplace wellness program such as stress management, weight management, smoking cessation, nutrition education, and recovery from substance abuse.

Regular physical activity plays an important role in all aspects of wellness. Exercise is considered the foundation of any wellness program. As a stress management tool it is as important as individual relaxation training.

It is widely known that weight management is more successful if regular exercise is incorporated.

Smokers frequently report the desire to smoke less when exercising regularly, and those involved in smoking cessation programs find that routine exercise reduces their perceived nicotine needs and smoking desires.

Recovering alcoholics and other substance abuse addicts experience a greater degree of success in the recovery process when regular exercise is included in their program.

Notwithstanding the accepted need for a strong fitness foundation is any workplace wellness program, these activities must be well thought out. Certain activities can be very intimidating to the "unfit" employee population. All levels of fitness must be addressed in the program's development in order to ensure success.

Variety is essential, and it should be expected that initial voluntary participation may be small. The physical activities chosen should be convenient for employees to participate in, and whenever possible management should show their commitment by allotting some work time for program participation. However, if activities are only to be offered after work hours, then child care services are strongly recommended.

As has been mentioned, in order for an individual to attain a state of optimum well being, all components of wellness must be addressed. However, it is important to note that if physiological changes are expected to take place in the form of lowering, eliminating, or preventing health risk factors, an individual must participate in some form of physical exercise.

Some individuals are motivated by the physical or aesthetic benefits of exercise. This can be a positive motivator for many people. However, there is a tendency for unfit or overweight individuals to make unrealistic comparisons with others and subsequently become very discouraged by their own progress. Again, individualization of goals and activities is essential to ensure that every participant remains motivated.

Participants should be reminded on a regular basis to direct their exercise efforts toward reducing health risk factors in order to optimize health and well-being. By focusing on specific risk factors a wellness program can motivate all participants to realize success without comparing themselves unrealistically with others.

Goal setting in fitness programming is highly successful when structured around reducing health risks. This allows participants with a negative body image or low self esteem to direct their attention and efforts toward more positive results.

ON-SITE AND OFF-SITE OPTIONS

Providing on-site fitness programs does not necessarily require a large capital expenditure in the form of a weight training facility or gymnasium. In fact development of a fitness facility is discouraged in the early stages of a wellness program. If the need and interest exists after a wellness program is established it can be considered at that time. The capital required to build facilities is more wisely allocated to other areas of the program during its initial stages.

It is, however, important in both the initial and long term stages of program development to offer the widest variety of fitness activity options as feasible within the program's budget. It may be necessary to include both on-site and off-site activities.

There will be some participants who will be motivated to exercise if they are provided an opportunity to do so before leaving the workplace to go home. Again, consideration should be given to allotting a share of work time for this activity, for example, a half hour of work time in an hour-long exercise period.

There will also be a number of participants who will have no desire to participate in any activities at the workplace. These employees should be provided an off-site option for engaging in the appropriate exercise activity.

Providing only on-site activities will limit the total number of employees who will be willing to participate in the wellness program. Utilization of community resources is essential in balancing options between on-site and off-site activities.

ON-SITE OPTIONS

Fitness Skill Classes

This type of class is an essential part of any fitness program. Many wellness program participants will have had very little or perhaps no experience in actually performing exercise. This will be especially true for the female population over 50 years of age. Those who have had experience with exercise activities have most likely had no direction or instruction. They are often misinformed about what is the appropriate exercise activity for them to participate in, as well as in regard to the intensity and the duration of performance.

Male employees, regardless of age, have most likely been exposed to physical fitness in athletics as an adolescent or in the military. Both experiences may have been less than ideal and resulted in negative attitudes toward routine physical exercise.

A skill building class should be considered a pre-exercise activity. All participants should receive some fitness training prior to performing their chosen exercise activities. This training can be incorporated in each activity offered, or stand alone as a separate class.

The content of the skill building class should include:

Exercise heart rates:

- Teaching participants how to monitor their own heart rate during exercise
- The importance of heart rate during exercise
- How to determine their appropriate exercise heart rate
- Appropriate and effective heart rate ranges during exercise
- The correlation between heart rate and cardiovascular condition
- The correlation between exercise heart rate and weight management
- The difference between aerobic and anaerobic energy systems

Determining appropriate exercise—duration, intensity, frequency:

- Teaching participants how hard to perform a specific exercise
- How intensity determines whether the exercise is aerobic or anaerobic
- How long to perform specific exercise activities
- How often to perform specific exercise activities in order to accomplish specific health and fitness goals or address particular risk factors which were identified in the needs assessment portion the wellness program.

Proper body alignment and mechanics:

- How to recognize good posture as it relates to exercise technique
- How to correct postural abnormalities modifiable by proper muscle development
- How to modify exercise movements to accommodate specific musculoskeletal limitations due to injury or genetic proclivity

- How to develop and balance opposing muscle groups for structural balance

Exercise safety and injury prevention:

- How to perform specific exercise activities in a safe and effective manner
- How to choose a safe exercise environment, e.g., effect of terrain type on walking, jogging, running regimes
- How to dress appropriately for exercise in various environments
- How to recognize and prevent heat exhaustion and hypothermia
- How to deal with outside extraneous factors such as traffic, air pollution, noise
- How to chose appropriate exercise equipment for various types of exercise regimes

Motivational techniques:

- How to stay self-motivated through enhanced physical well-being states
- Explanation of employer provided incentives for goal accomplishment
- How to choose an exercise partner to complement and aid in goal acquisition
- What to expect in regard to instructor contact and assistance
- Encouragement of participant feedback through the workplace wellness coordinator
- What regular testing, monitoring and evaluation will be provided to track progress and provide information for program modification

Exercise Leaders

An exercise skill building class should be taught by a qualified fitness instructor. However, interested employees should be designated as contact personal to provide information and to help motivate other employees to participate. Highly motivated employees can be utilized as instructor assistants.

It is important to note that skill building classes are not commonly offered at community facilities. The corporate wellness director must coordinate skill building classes with employee utilization of off-site facilities.

Walking Programs

Walking programs are frequently the most popular of the exercise activities, and therefore the most highly attended by novice participants. There are two primary reasons for this attendance preference:

1. Walking is something that most everyone can do with ease.
2. Walking can be done anywhere, at anytime.

These two factors make walking very convenient, and an excellent startup exercise. However, the strategy of a walking program should be well thought out to maximize participation.

Using responses from interest surveys conducted during phase 1 of the wellness program, wellness coordinators should determine specific walking interest, that is, how many employees currently walk for health reasons and how many would begin walking if a program were offered.

Once these interest questions have been answered the coordinator should analyze the work environment and surrounding terrain to determine appropriate areas for walking.

Depending on the workplace location, walking areas could range from a designated floor of an office complex to nature trails around the workplace. Walking trails can include parks, city blocks, shopping malls, or trails which are established around the perimeters of enclosed work areas. In more rural areas, nature trails can be developed, and perhaps opened to the community as a public relations gesture. Program coordinators should make contact with those employees already walking on or near the worksite to gather information and ideas for designated walking areas.

Once appropriate routes for walking have been established, wellness coordinators should begin to publicize the walking program to all employees. Planning a kickoff date can be very effective. It is important to note that upper management participation is highly recommended.

Walking programs also create an opportunity to break through some of the normal communication barriers that are due to management and staff stratification. Improving communication is always a significant step toward developing higher productivity and increased profits.

Once a walking program is begun it should be considered an ongoing activity.

A skill building class should be conducted even for an exercise program as basic as walking. All the components listed in the exercise skill building class are necessary in order for fitness walkers to perform in a safe and effective manner, either individually or as a group.

Fitness walkers should be encouraged to take advantage of opportunities away from work to continue their walking programs on weekends and holidays. Family involvement should also be encouraged.

If individuals are participating in a walking program to reduce health risk factors, these factors should be monitored on a regular basis to ensure safe progress. It should be noted that any employee participating in a fitness walking program should do so only after a health and fitness screening, and after receiving appropriate medical clearance if necessary.

Individuals participating in a walking program to manage their weight should participate in a specific weight management program as well. The need for additional participation is important because single-level weight management regimes will reach a point of diminishing returns. When this occurs participants become discouraged and fall out of the program. Weight management, in particular where weight loss is desired, requires specific monitoring and educational techniques to ensure success. Additionally, exercise programs must be adjusted frequently in accordance with the participants' progress.

Although walking programs are very basic, program coordinators need to monitor and evaluate the results of the program. Such evaluation should include:

- Number of participants. Is participation steady, or does it fluctuate? Can the low spots be improved with the addition of incentives?
- Percentage of employee population participating—feedback as to reasons for participation, nonparticipation, or dropout.

- Perception of effectiveness by participants. Changes in effectiveness perception can act as a tripwire for the introduction of the next level of fitness activity where a walking program is utilized as a startup tool. As participants become stronger and more fit, walking alone may not satisfy their fitness needs or desires.
- Changes in individual health risk factors.

Running Programs

Usually, employees who express an interest in running or jogging already participate in such a program on their own, or have done so in the recent past. These individuals are usually fitness self-motivated and are interested in educating themselves further in regard to an activity that they already participate in, increasing available time for their participation, or sharing their activity with other employees who have similar interests.

Any running or jogging program should be considered an advanced level of activity, and should be offered to the more fit employee. Should an employee with average or below average fitness levels express an interest in running or jogging, it is recommended that they begin their training in the walking program and build endurance and strength before engaging in a company sponsored running or jogging program.

Exercise skill building classes associated with running or jogging should be specific and address footwear, clothing, terrain, injury prevention, weather, hypothermia, hyperthermia, intensity, flexibility and stretching, proper warmup, nutrition, and body mechanics, along with the factors mentioned above under skill building classes.

Running or jogging can provide maximum benefits very quickly, but can also lead to serious injury or increased health risk if not pursued and monitored appropriately. Only qualified fitness instructors with running experience should teach this skill building class.

Additionally, participants should be encouraged to read running related journals and magazines to aid in continued motivation, and for continuing education.

The employer can gain a significant public relations advantage, as well as enhanced employee participation motivation, by sponsoring community fun run events.

As a group runners tend to be dedicated individuals and can be very helpful to wellness directors and coordinators in motivating coworkers to participate in wellness activities.

As with walking, running or jogging programs should be considered an ongoing activity. Coordinators should maintain periodic contact with participants to monitor progress and to evaluate the activity.

Stretching

Stretching for flexibility is very important and should be included in every exercise activity. However, it can also stand alone as an activity in itself.

When stretching for a specific activity, such as running, walking, or aerobic dance, the stretching exercises tend to be specific to the major muscle groups involved in the activity.

A standalone stretching class, on the other hand, allows time for a more total body stretch. A stretching class is an activity that will attract individuals who are not yet ready to commit to a full exercise regime, or those individuals who have specific limitations that preclude them from exercising more vigorously. The older population of the workforce will be more attracted to a stretching program than to other types of activity. Obese individuals will also often take their first step toward exercise through a stretching program.

Stretching may be performed with two objectives in mind:

1. Stretching for flexibility
2. Stretching for relaxation

Both goals can be addressed in a class setting, and subsequently pursued individually or in group work. However, normally only one type of stretching goal is emphasized in any given class.

Stretching classes should be taught by a qualified instructor. Incorrect stretching technique may lead to injury or the increased chance of injury during other fitness exercises or sports. A stretching instructor must be well versed in anatomy and body mechanics, and should also have some knowledge of and experience in working with special populations such as older participants and the obese.

Yoga is sometimes an option offered in a stretching program. Yoga can be a very effective discipline, but some yoga moves are radical and may be inappropriate for certain individuals. The wellness director should make certain that all yoga moves are discussed and agreed upon before class inception.

Also, yoga is considered to be a religious act by some, who may be offended by its incorporation in a workplace wellness program. A response survey addressing this possible reaction should be completed prior to developing a yoga program. A written statement as to the nonreligious purpose of the yoga program would be advisable as well.

In addition to a stretching class, an effective way to involve the majority of the employee population in stretching is to organize groups by department, and design specific stretches related to their work duties. Incorporating such stretching into daily work routines can reduce stress, reduce repetitive motion injuries, and improve productivity.

Examples of such work specific stretching include:

- Desk stretches for clerical personnel
- Back stretches for construction and line manufacturing personnel

Work specific warmup and stretching activities have been shown statistically to reduce the frequency and cost of workers' compensation claims.

Job related stretching programs should be developed and delivered by qualified personnel with working knowledge of anatomy and body mechanics, and with experience in teaching stretching in the work setting.

Aerobic Dance

Aerobic dance is frequently a component of on-site fitness activities. It requires little capital expenditure for facilities. Certified aerobic dance instructors are abundant in most areas of the world and can be contracted to teach on-site classes at reasonable cost. A multilevel, low impact aerobic dance class can be an effective exercise activity for employees at all levels of fitness. Offered in conjunction with other on-site fitness activities, such a class is a good addition to a workplace wellness program. However, if aerobic dance is offered as the only on-site program, participation will be limited.

Aerobic dance classes tend to attract a female employee population with an average to above average fitness level. Males typically do not participate in this type of fitness activity. A percentage of fit males will participate, but this percentage will be very small.

As with all exercise programs, coordinators should monitor and evaluate the effectiveness of the program on an ongoing basis.

Aerobic dance raises the question of coeducational participation: should it be encouraged or discouraged?

The question of coeducational participation need not be of major concern. However, to avoid the possibility of sex discrimination all activities should be open to male and female participants. A variety of fitness activity options should be offered.

As noted above, some activities will attract predominantly female participants, while others will attract predominantly males. Certainly an employer should avoid offering only one activity which tends to attract only one segment of the employee population, be it younger, older, male, or female.

However, where preliminary skills building classes are concerned separation by sex, age, and physical condition is a good idea. Because of physiological differences, the frequency, proper body alignment, intensity, and duration of any given activity may be substantially different for each of these groups.

Where skill building training is incorporated into participation in a specific exercise program, body mechanics will have to be taught separately to all segments of the participating population before the actual exercise period.

For example, flexibility goals in leg stretches specific for running, aerobics, or walking may be different for different sectors of the population. Generally, young female participants will show much greater flexibility in certain leg stretches than young males, older females, or older males.

Regardless of how body mechanics skills are taught, the instructor must give consideration to these differences.

The advantage of individual segregated skills classes is that the entire time can be devoted to the needs of a specific employee population grouping, whereas in multifaceted groupings one group is addressed while the others wait their turn.

Circuit Training

A circuit training classes is an organized, group led exercise activity that integrates cardiovascular conditioning and resistance training into a fitness class format. This approach to fitness is very efficient in terms of the amount of time spent to accomplish specific goals. It is appropriate for individuals at all levels of fitness, and is easily modified for individuals with specific physiological and/or musculoskeletal limitations.

Males are attracted to this type of exercise activity because of the strength training component and because the aerobic components tend to be more simplified than an aerobic dance class.

Females may initially be intimidated by some of the equipment used, but with proper introduction and training are quickly won over to the program because of the rapidity of goal attainment versus the time commitment required.

An on-site circuit training facility may require significant capital outlay if strength training and resistance equipment is utilized. This type of training facility is for the advanced workplace wellness program, where management is thoroughly committed and where on-site space is available to build a full line fitness facility. Imagination and capital are the only limiting factors in the development of such a facility.

Fitness center consultants should be contacted for input in designing and outfitting a workplace fitness center. Acquisition of quality high use, low maintenance equipment is of primary importance. The order in which equipment is purchased is also important. Certain specific equipment should be purchased first, to build a solid foundation; then, as the facility develops, other equipment can be added to round out the exercise options offered.

Fitness instructors can be contracted to teach classes. The content is fairly new in the fitness industry, and refinement of courses and equipment is constantly evolving. Look for instructors with strong experience; class content should be their responsibility.

Aerobic Fitness Facility

An inexpensive approach to an on-site fitness facility is an aerobic equipment room. Aerobic equipment includes stationary bicycles, stair climbers, rowing machines, and cross country ski machines. For employers fully committed to a workplace wellness program, an aerobic facility can evolve, over time, into a more diverse aerobic and weight training facility. However, if that is the ultimate goal then fitness center design consultants should be brought in during the initial design phase to ensure that all available space is most efficiently utilized.

An aerobic facility generally requires minimal skills building instruction. It is an exercise center that can be used by individuals who prefer working out alone. By utilizing a variety of light resistance equipment an individual may condition all major muscle groups.

The facility may also be used for short workouts during breaks, or for shift workers who are not normally on site for other wellness program options.

Additionally, a portion of the center's space can be reserved for aerobic dance classes.

OFF-SITE FITNESS PROGRAMS

There are both advantages and disadvantages to off-site fitness programs. The most obvious advantage is cost reduction. However, as we shall see, the apparent cost reduction may be illusory at best.

Many communities have facilities, generally within the parks and recreation department, which have fitness programs that can be utilized by the workplace population for implementation of both the fitness training and the recreational portions of a workplace wellness program.

The private health club is another off-site alternative. However, community facilities are usually less expensive than private facilities of the same or better quality.

Either option offers an opportunity for the wellness director to negotiate a group rate for the use of the facility by wellness program participants.

The primary disadvantage of off-site fitness programs is the loss of supervisory and management control over the content and quality of the program. Additionally, it is difficult to track results, do followup assessments, and otherwise utilize the facility as effectively as when an on-site facility is available.

Community programs are seldom geared to provide the personalized instruction required by the employer's needs assessment. They tend to suffer from budgetary constraints and often hire relatively inexperienced fitness instructors, who are willing to exchange a low salary for the chance to gain experience. This lack of experience may also reduce the effectiveness of the employer's program.

Consequently, the wellness director may find that an inordinate amount of time is required to monitor the progress and quality of the outside program. Such monitoring requires a significant time commitment and may in fact offset the savings anticipated by the use of an off-site facility.

While off-site fitness training may not be a viable option for the majority of the workplace population, it is certainly a viable option for employees who are self-motivated toward fitness maintenance prior to the inception of the workplace program, and for those employees who simply prefer to work out away from their place of employment.

Also, private health clubs usually provide a diversity of equipment and fitness activity. Unfortunately, when it comes to instruction, private clubs suffer from the same drawbacks as community facilities.

Workplace populations as a whole need to be properly trained in the goals and techniques of fitness attainment before being turned loose to proceed on their own in an unsupervised setting.

Of course, neither of the above off-site options would necessarily preclude the possibility of using qualified instructors retained by the employer to train employees at the off-site location. The wellness director will need to make the necessary arrangements and be prepared to hold the facility harmless for injuries resulting from any instruction provided by the employer's trainer.

The initial needs assessment can be designed to determine what percentage of the population would prefer an on-site facility, and what percentage would prefer an off-site facility. This information will be of great value to the employer in designing the fitness portion of the wellness program, and in determining which of the two options is best for the workplace population. A combination of both options is most often the best choice.

RECREATIONAL PROGRAMS

Recreational programs, i.e., team sports participation and some individual competition sports such as bicycling, swimming, or running, can be a very beneficial adjunct to an employer's overall program.

Recreational programs traditionally are of great interest to a large number of employees when they are included on interest surveys completed during the design stage of a wellness program, especially when the activities offered correspond to those which a majority of employees have already participated in. Volley ball, softball, and basketball always seem to be popular.

Offering a recreational program can attract those employees who are reluctant to get involved in a fitness or strength training program. Unfamiliar activities always seem a bit intimidating.

An effective way to utilize the recreation program is to develop opposing teams and create tournaments within the workplace population. In some cases these competitions can take place between teams fielded by other employers within the community, or even within a fairly large geographical area.

Again, employees should not participate in recreational team activities without appropriate fitness assessment and training. Therefore, the incentive of being a team member and participating in tournaments can be a successful motivating factor in encouraging otherwise recalcitrant employees to participate in the employer's fitness training program.

Team sports, when utilized appropriately within the total workplace wellness program, can help lower health risk factors as effectively as other more traditional activities of the wellness program.

Team activities can affect morale positively and promote a strong esprit de corps among the workplace population.

There are, however, liabilities associated with team sports, as there are with any activities associated with a wellness program. The employer needs to be aware that an injury suffered by an employee while participating in an employer sponsored or sanctioned activity may be considered compensable under workers' compensation law.

This liability can be mitigated through a number of employer strategies. The first and most important strategy is the careful design and management of the wellness program itself, making certain that quality training is provided throughout all phases of the program.

Team sports should be offered as an option only after appropriate training, and should only be scheduled after working hours. Whether or not they occur on site makes little or no difference from a workers' compensation standpoint.

Employees should be informed that participation in the team sports is absolutely voluntary and primarily for their own enjoyment and recreation.

Workers' compensation law does not require that an employee

seek compensation for a team sport related injury exclusively through the workers' compensation system. In fact, in most states the law itself does not address the question at all. However, in general, administrative law judges have ruled in favor of employee compensation within the system when alternative treatment was not available to the injured employee.

It therefore behooves the employer interested in sponsoring team sports to purchase an accident insurance policy which covers participating team members for injury arising out of, and only out of, team related activity, including practice and tournament play. The policy should require no employee deductable or coinsurance. The employee will therefore not need to access the workers' compensation system.

The rationale for a separate team participation injury insurance policy is that such injuries will not then affect the employer's workers' compensation experience rating, which in turn controls the overall cost of workers' compensation insurance.

Team participation insurance policies are relatively inexpensive because they cover a limited number of people for a specified period of time.

Additional liabilities can be incurred in association with team sports; it is not within the scope of this work to discuss these risks in their entirety. We recommend that a thorough exploration of the risks be conducted with input from the employer's legal council and insurance provider. Potential liabilities are a fact of life, but they generally do not prohibit participation in team sports when the risk is duly assessed and the employer is appropriately insured.

Another competitive recreational approach which should be considered in conjunction with or as an alternative to team sports is individual competition within a tournament of several events, i.e., an employee Olympics style competition. This type of activity tends to attract the more individualistic employee who would not participate in team sports, but would like to have an opportunity to compete and win in individual competition.

Chapter 6

Smoking Cessation and Substance Abuse Programs

SMOKING CESSATION

A smoking cessation and tobacco education program should include the entire employee population. The program should not direct its attention only to those workers who currently smoke. Tobacco issues should be addressed as a wellness issue of the general workplace population.

The Environmental Protection Agency has designated second-hand tobacco smoke as a class A carcinogen. Second-hand tobacco smoke is that smoke which pollutes the immediate atmosphere surrounding a smoker and is breathed in by other smokers or non-smokers. Second-hand tobacco smoke is considered by the EPA to be as much as 30% more carcinogenic to nonsmokers than the direct smoke inhaled by the person smoking.

The goal in a smoking cessation and tobacco education program is to:

1. Encourage current smokers to quit smoking or using tobacco
2. Discourage nonsmokers from taking up the habit
3. Secondhand smoke education

In order for the program to be completely successful, and to reduce the risk of workers' compensation claims arising from a tobacco smoke polluted environment, employers are strongly encouraged to develop and implement a smokefree workplace policy. An implied liability exists in regard to workers' compensation cumulative illness claims which may arise out of a nonsmoker's illness as the result of cumulative exposure to second-hand smoke. Workers' compensation claims of this type have already been successfully filed.

A variety of studies and surveys have shown that the majority of those who currently smoke would like to quit permanently. Having both a smoking cessation and tobacco education program available at work as well as a smokefree workplace environment will greatly improve the chances of success for those individuals who do make the decision to quit the habit of smoking or chewing tobacco.

It is strongly recommended that the wellness director work closely with program implementors to create a positive workplace environment for participation in a smoking cessation and tobacco education program. There is the tendency for smokers to feel intimidated by smokefree workplace policies, and in some instances to feel that they have been discriminated against by the implementation of such policies. These feelings may surface in spite of the fact that the smoker truly desires to quit what is logically perceived to be an unhealthy, life threatening habit.

Nicotine, one of the primary components of tobacco smoke, is considered to be a highly addictive substance. Smoking produces significant cravings which are both physical and psychological. As a result, many people are able to quit more successfully with some form of external encouragement and support.

Certain segments of the population are more inclined to take up the habit of smoking. These population segments are under very heavy peer and marketing pressure to ignore the known adverse health effects of smoking. Hence a social reward is presented which in the short term may be stronger than the long term reward of nonparticipation.

Again, a positive smokefree workplace policy, with nonsmoking co-worker support, can be very effective in reducing external social pressures. After all it must be remembered that most workers spend at least one-third of their daily lives at the workplace.

Notwithstanding policy decisions to establish a smokefree workplace, participation in smoking cessation programs should be voluntary. The program should be well publicized and presented in a manner that is nonintimidating. Every attempt should be made to make the smoker feel comfortable about participating in the program.

The preliminary stages of a tobacco education program can be used to determine the best approach to encourage smokers to participate in the smoking cessation program. To ensure the greatest possible success, smoking cessation programs should be offered on work time, with employees receiving their normal pay for participation.

Although the employer will be absorbing a significant labor cost by such an approach, the return in success rate will make the investment worthwhile.

Consider the following information. A conservative estimate of the increased health care costs to an employer by tobacco smoking employees is in the range of $300 to $600 per employee per year. These costs do not take into consideration the loss in productivity caused by employees on sick leave due to smoking related ailments. And without a doubt smokers have a statistically greater rate of absenteeism than nonsmokers.

Participation in the smoking cessation program should be made as convenient as possible for the employee. This will eliminate one of the most common excuses smokers have for nonparticipation in a smoking cessation program, i.e., conflict with other time commitments. Nicotine addiction stimulates creative reasons for nonparticipation. Successful cessation programs anticipate and eliminate as many of the nonparticipation excuses as possible.

The most convenient place for participation is the work site. Off-site programs have greatly reduced participation rates. While success within the population of attendees is not reduced by an off-site program, the overall success rate within the workplace population is greatly reduced.

There are a number of standalone smoking cessation programs available on the market today.

Staff personnel can be very effective leaders, albeit some professional training may be required prior to program implementation. Regardless of who is chosen as leader, his or her ability to relate to the cravings experienced by individuals attempting to quit the smoking habit, and a leader's genuine desire to help these people, are of significant importance. This does not mean that the leader should necessarily be a reformed smoker. Additionally, of primary importance is an assessment as to the willingness of identified personnel to act as program leaders.

There is a natural tendency to measure the success of a smoking cessation program by the percentage of people who successfully quit smoking against those who return to smoking after completion of the program. Certainly this is valid. It should be remembered that no smoking cessation program will be 100% effective the first time it is introduced. However, with continued positive peer support at the work place, and participation in remedial cessation groups, 100% success is very feasible.

The first 72 hours of a cessation program are the most critical. This is the time when most individuals will feel the strongest cravings for nicotine and the strongest desires to continue smoking. Once that time has passed the individual has an increasingly better chance of complete success.

Interestingly, another critical period may be reached 3 to 5 weeks after successful smoking cessation. Here the individual begins to feel so good, as the result of smoking cessation, that there is desire to return to the old habit almost as a celebration of their success. It is at this critical juncture that other aspects of the company's wellness program, i.e., fitness, team sports, nonsmoking support groups, etc., are most important to continued abstinence.

Followup on a smoking cessation program is important to assess overall success rates and to determine when another program should be initiated. However, it should be remembered that there is very little that an individual instructor can do to ensure long term successful smoking cessation through periodic followup. It is the overall effect of individual commitment, peer support, and other positive elements of the workplace wellness program which will ensure long term success.

ALCOHOL AND CONTROLLED SUBSTANCE REHABILITATION

The escalating problem of work related injury and illness associated with the use of alcohol and controlled substances in the workplace has risen to totally unacceptable levels for many employers.

It is important for all employers to make every effort reasonably within their budgetary power to address and eliminate substance abuse from the workforce.

The first step in this process is to develop a zero tolerance substance abuse policy at the workplace. The substance abuse policy should clearly define the employer's position and clearly state the disciplinary action which will be taken in the event an employee is found to be under the influence of alcohol or controlled substances on the job.

At the time of this writing, Federal guidelines in regard to employer substance abuse policies are unclear. National policy indicates that an all-out effort is being made to eliminate controlled substances from our society, yet specific guidelines in regard to employer responsibilities do not exist. The only sector of our economy which has developed firm substance abuse guidelines is the transportation industry.

The courts have generally ruled, on privacy grounds, that employers do not have the right to test employees at random unless the duties the employees perform create an issue of public safety. It is therefore suggested that employers consult with legal counsel specializing in labor law when developing a comprehensive substance abuse policy.

Many employers find that dealing with substance abuse strictly as a performance issue is the best solution, disciplining employees for poor performance, with termination as the ultimate disciplinary tool.

While this position is prudent it does not resolve the long term issue, nor dies it take into consideration action to be taken when an employee voluntarily divulges a substance abuse problem and requests rehabilitation assistance from the employer.

We feel that a dual approach to addressing the substance abuse question is prudent. Definitely a stated zero tolerance policy is required. An example of such a policy is given below.

WARNING: DO NOT IMPLEMENT THE FOLLOWING POLICY WITHOUT SEEKING THE ADVICE OF AN ATTORNEY WELL VERSED IN LABOR LAW, AND LOCAL DISTRICT RULINGS OF THE EQUAL EMPLOYMENT OPPORTUNITY COMMISSION. THE FOLLOWING IS MEANT AS A SUGGESTED GUIDELINE ONLY.

> An employee who works under the influence of controlled substances, or alcohol is a danger to themselves and their fellow workers. Such an exposure cannot be tolerated.
>
> (Company name) does not tolerate the use of controlled substances or alcohol abuse. (Company name) will not allow any employee to work while under the influence of either alcohol or a controlled substance.
>
> (Company name) encourages employees to participate in substance abuse

rehabilitation programs voluntarily. Successful participation will reflect positively, not adversely, on the employee's future advancement opportunities.

(Signature of Company President)
(Date)

The establishment of an employee drug screening policy in regard to the existing workforce should be deferred until definite E.E.O.C. guidelines are promulgated and supported by court rulings. However, the establishment of a drug screening program in conjunction with pre-employment physical examinations may be developed under the advice of appropriate legal counsel.

By eliminating or at least reducing the chance of hiring an employee with a substance abuse problem, the employer will begin to fill the workforce with nonabusers who will support the employer's zero tolerance policy and who will over time, through peer pressure, drive substance abusing workers into the employer's rehabilitation program or out of the workplace voluntarily.

Another area where drug screening may be successfully used is in association with work related accident investigations. Again with the advice of appropriate legal counsel, the employer can develop a drug screening policy which will test all employees involved in work related accidents that result in injury beyond minor first aid, or injury to a third party even if minor, or result in near injury to a third party, or result in substantial material damage to property or equipment. Substantial damage should be defined by a minimum dollar figure.

A definite policy of discipline must also be stated. At present the E.E.O.C. encourages employers to establish, or at least refer employees identified as substance abusers, to substance abuse rehabilitation programs and subsequently reintegrate the rehabilitated employee into the employer's workforce. We are in agreement with this approach where it is feasible within the employer's operation.

A substance abuse rehabilitation program should always be conducted under the supervision of a substance abuse rehabilitation professional. Under no circumstances should an employer attempt to self-administer such a rehabilitation program unless staff professionals have been hired or retained.

In addition, the overall workplace wellness program can be very effective in reducing employee involvement with controlled substances and alcohol.

It must be remembered that of the vast number of people involved in the use of controlled substances and alcohol, the majority are not so adversely affected by their use of controlled substances that it is apparent in their work performance. Many users are relatively well disciplined people who consider moderate drug use no more physically debilitating than moderate alcohol use. Unfortunately, these individuals contribute to the overall problem by economically supporting the illicit trade of recreational drugs and socially encouraging peers who may not have equal usage discipline.

The best approach to discouraging the moderate user is subtle peer pressure from nonusers, coupled with educational programs which inform the moderate user of the serious economic and physical drawbacks to continued use. This approach is best accomplished under the auspices of the overall workplace wellness program.

Before leaving this subject we must address the criminal aspects of controlled substance trade and use. Under no circumstances should an employer allow trade in controlled substances to go unchecked within the workplace. Unfortunately, many employers simply turn a blind eye to these transactions either because of a lack of resolve, or for lack of knowledge in regard to how the problem can be addressed. Allowing trade in controlled substances to continue unabated will ultimately defeat the success of the employer's substance abuse efforts.

Notwithstanding these admonitions no employer should attempt to deal with controlled substance trade unilaterally. This is an area where professional law enforcement assistance is needed. Employers should establish contact with local, or Federal law enforcement personnel and develop a liaison program which addresses their individual controlled substance trade concerns.

Chapter 7

Weight Management and Nutrition Education

A weight management program should be offered on a continuous basis to ensure long term success. In order for an individual to attain long term success in weight management, lifetime habits of eating and exercise participation must be identified.

Successful long term weight management cannot be attained by modification of eating habits alone. Initial weight loss is not sustained once the individual returns to former eating habits. This type of weight management plan is, therefore, very discouraging and unsuccessful in the long run.

In order to ensure success, a weight management program must deemphasize the weight loss aspects of diet, and emphasize the overall health benefits of appropriate body composition. In addition, changes in diet must be made gradually to ensure that the individuals participating in the program do not feel deprived of their favorite foods. Feelings of deprivation create cravings which in turn hinder or completely obstruct successful attainment of program goals.

Education in the basics of nutrition and in the application of nutritional guidelines is necessary to the success of the program.

It is most effective to begin a weight management program by offering a general nutrition class to the total workplace population. This approach accomplishes two goals: (1) It avoids the risk of generating discriminatory animosity among obese employees who feel that they have been singled out due to their overweight condition. And (2) it addresses the needs of members of the workplace population who may have diet related risk factors which are not apparent through simple observation of their appearance.

Nonobese individuals may have the same risk factors as obese individuals, including elevated cholesterol and/or triglycerides and high blood pressure. Appearance does not necessarily indicate nutritional knowledge or health status.

The general nutritional education component of a successful weight management program will reach all individuals who have the need and/or desire to improve their nutritional awareness.

Immediately following general nutritional education is the weight management portion of the program, which is specifically designed for the overweight component of the employee population. Ideally, this portion of the program is entered so smoothly that overweight individuals do not realize that they alone are being addressed. Gradually, as the material presented becomes less applicable to nonobese individuals, their attendance at the classes will begin to drop off, ultimately leaving only those in greatest need of assistance in weight management.

The focus of weight management is to reduce diet-related health risks, which include an increased chance of heart disease, diabetes, and some cancers whose risk of occurrence is related to a high fat, low fiber diet.

It is the overall goal of a workplace wellness program to modify employee lifestyle and offer alternatives to unhealthful habits in an effort to reduce health care claims. Such a reduction improves both the health care benefits cost profile and worker productivity, which in turn increases employer profits.

When overall health factors are emphasized in a weight management program, the focus for participating individuals is away from loss of weight and toward improved health and well being. This is the primary key to long term success. Aesthetic consideration is a short term reward which should not be overemphasized.

If the program focuses on the benefits of health and well-being through weight management, the importance of appearance to self-esteem becomes no more than a secondary consideration. Transient changes in personal relationships and social interactions will, therefore, be less likely to drive the participating individual out of the program. Self-esteem will be enhanced by successful attainment of program goals even if appearance changes are not immediately recognized. As program goals are met, concomitant appearance changes may occur which will enhance self-esteem.

Self-esteem is extremely important to an individual's success, and as such directly relates to productivity and esprit de corps. Before one can feel good about being a part of a social unit, one must have a certain level of self-esteem. The greater the self-esteem the more successful one will be in all aspects of life. The benefits of enhanced self-esteem within the workplace population translates directly into higher productivity and profits for the employer.

Instructor guided support groups are essential to an effective weight management program. Developing such groups among interested employees is a way to facilitate the daily assistance that is required during the initial stages of a weight management program. Leaders from within the workplace population can be developed by the instructor. To be most effective these group leaders should be chosen from among the population of employees who have successfully altered their former unhealthy lifestyle habits, and who are successfully managing their weight.

There is an abundance of misinformation and misconceptions concerning weight management and weight reduction. It is, therefore, important that the workplace population receive accurate information presented with sufficient frequency to dispel the effects of erroneous advertising and factual misunderstanding. Again support groups can be very effective in this regard.

Support groups can provide the long term assistance necessary to ensure success. They provide an arena for interaction which can overcome outside influences. New fads, diets, and regimes can be discussed and placed in their proper perspective.

An effective weight management program will generate weight loss through lifestyle changes, and behavioral modification related to dietary habits. It will also provide information to individuals in a manner that will allow each participant to apply specific techniques to his or her particular needs.

It is strongly recommended that artificial dietary approaches which involve drastic dietary changes be strictly avoided. Such regimes would include diets composed of food substitutes, diet drink concoctions, and nonnutritional bulk drinks which replace nutritious food with fiber fillers.

These programs result in sudden and in some instances drastic temporary weight loss. Weight is regained as soon as the participating individual returns to a normal diet. Of greater concern is the long term deleterious effect of such diets on those individuals with superior will power who remain on these regimes for extended periods of time.

Successful weight management requires modification of eating habits. This modification must necessarily take place gradually, over a relatively long period of time.

Once a weight management program has been implemented it is important for the employer to signal its support of the program through subtle environmental signals. An obvious example is the content of on-site food vending machines.

Vending machine content should be analyzed and evaluated in regard to its nutritional content, and gradually modified to reflect the changing interests and desires of the workplace population. This can be facilitated through utilization of interest surveys placed near the machines themselves.

Drastically altering vending machine content is not recommended. Such instant changes will result in revenue loss and disgruntled employees. However, there are nutritional differences between the various packaged snack foods available on the market. Moving gradually toward healthier, less fatty foods is strongly encouraged. Eventually, a pure nutritious food snack vending machine can be installed as an alternative, and the vending machine sales monitored. Gradually more and more of the workplace population will purchase foods from the "healthier" machine.

However, it is not anticipated that the less nutritious machine will ever be completely eliminated. Still, even within the category of "food for entertainment" there are healthier alternatives, particularly in the area of reduced saturated fats and salt content.

Of all the opportunities available to the employer to show support for a weight management effort, the on-site cafeteria has the greatest potential. Here the employer has the opportunity to gradually improve the nutritional content of the food consumed by the employee population during the work day. Alterations in the food offered can be linked to the progress of the overall weight management program, thereby encouraging employees to apply what has been learned in weight management classes. Such changes do not have to be drastic or expensive.

Offering choices to the workforce which are desirable and palatable, but lower in fat percentage, higher in complex carbohydrates, and which provide increased energy through ease of metabolism, will support weight management efforts and improve productivity over the long term.

Stable blood sugar levels are needed to maintain consistent work energy levels and task attentiveness throughout the day. Quality meal selection is essential to stable blood sugar. Employees who indulge in high sugar snacks and foods to the exclusion of adequate dietary needs will experience wide swings in blood sugar levels. These swings can result in mood alterations which include the inability to concentrate, psychological depression, and profound fatigue. All of which may be significant factors in work related injuries.

Again, interest questionnaires will aid in determining the food preferences of the workplace population. As the population becomes better educated nutritionally, and as the weight management program begins to affect behavioral modification, food preferences will change accordingly.

Utilization of both cafeteria management and an outside dietician will aid in determining precisely how to alter and improve the nutritional quality of the meals offered.

An important consideration when choosing an outside consultant for development of the overall workplace wellness program is the consultant's expertise in nutritional education and weight management. It is important for program success to incorporate the appropriate fitness activities into a weight management program, and correspondingly incorporate nutritional education into the fitness and recreational aspects of the wellness program.

Weight management must have an interrelated exercise program in order to achieve long term success. Diet alone cannot effect long term change. Routine exercise is necessary to ensure long term metabolic changes within the body which enhance the ability of an individual to utilize lipid reserves as an energy source.

Stress management is also a factor in successful weight management. Often individuals respond to increased stress with a desire to overeat. Stress management as a part of the overall wellness program is therefore necessary to a successful weight management program.

Monitoring the success of the weight management program is important. However, the method of evaluation should not focus on total weight lost. The most valid measurement of success is that of body composition. All participants should have an initial health/fitness assessment before entering the program. The assessment should include body composition in relation to total body weight, deemphasizing weight as a measurement and substituting in its place the concept of stored body fat versus lean muscle mass. The goal of the program is to decrease the percentage of stored body fat while increasing the relative percentage of lean muscle mass.

Participants also need to be educated in regard to the realistic body composition they can expect to attain for their age and body type, to ensure that goals are not set which if not met would reduce self-esteem and hence continued participation in the program.

By focusing evaluation away from weight lost the instructor facilitates long term success for the individual by readjusting goals to those related to overall health and well being rather than the purely aesthetic considerations of weight management.

Nevertheless, a normal byproduct of weight management is overall body size reduction, which is clearly evidenced by the need to purchase smaller size clothing over a period of time. Because of society's heavy emphasis on appearance this downsizing often improves self-esteem. On the other hand, if it is delayed in its manifestation the participants will not become discouraged because they are not focusing on only that aspect of the program.

It is a good idea to monitor body composition on a quarterly basis. More frequent monitoring tends to focus on a perceived need for rapid fat loss, which in fact is not the goal of the program.

There should be consistency in the means of evaluation and the person doing the evaluation. There will always be a degree of human error associated with mechanical forms of body composition testing. Frequent rotation of the evaluator compounds the human error factor, which in turn erodes the success of the program.

Of the various methods of body composition evaluation, water immersion measurement (hydrostatic) is the most accurate. It is also highly impractical for most workplace wellness weight management programs. Skin fold measurement is more practical.

The importance of training the program participants to exercise regularly cannot be overstated. Regular exercise at an appropriate level of intensity enables an individual to utilize stored body fat as a primary exercise fuel source. This then becomes an extremely effective means of altering body composition.

Concomitantly training individuals to recognize and reduce fat in the diet while participating in appropriate exercise rapidly addresses the excess body fat stores and alters total body composition in a positive manner.

Since the ability to participate in exercise is greatly effected by one's total body composition, each individual must be placed in an exercise regime which corresponds to his or her needs and abilities. A generic exercise program for weight management is not always feasible. However, walking programs generally involve all participants at their individual levels of ability. As they progress in the program additional or alternative exercise may be introduced.

Flexibility in program design and consistency of participation and instruction are the key factors in program success.

Finally, it is important to organize the program in a way that allows participants to join at any time. There will always be some who take a "wait and see" attitude, joining the program only after observing peer success.

Therefore, it is advisable to offer frequent introductory sessions which are primarily of an educational nature. Individual participants may then be assigned to existing support groups which best fit their individual goals and needs. Many participants will have attempted weight management on their own with varying degrees of success. Placing them in appropriate support groups is important in order to avoid the likelihood of subsequent dropout.

As more and more members of the workplace population attain their individual goals total program momentum will be achieved, eventually encompassing the entire employee population.

Chapter 8

Stress Management

The development of stress management is beneficial and necessary for a complete and effective workplace wellness program.

Before attempting to establish a stress management module, the employer must understand what is meant by the term *stress*.

Stress, as we define it for wellness program purposes, is any force or influence which causes one to feel anxious about one's ability to continue to perform one's job duties in a satisfactory manner. By satisfactory we mean both in terms of performance quality and job satisfaction as perceived by the employee.

The term *perceived* is extremely important. Under normal circumstances it is not the stressor itself which adversely affects the employee, but rather it is the employee's perception of the stress which creates the problem. Therefore, a goal of any stress management program is the education of employees in regard to their perception of stress.

Regardless of the source of stress it should be remembered that any stress is unhealthy if it is perceived by the individual as negative and hence deleterious to their health.

Stress is always perceived as negative when the individual experiencing the stress has no control over the magnitude of the stressor or the duration of its influence. Any arrangement which gives an individual the opportunity to exert some degree of control over a given stressor will reduce its perceived negative effects.

Stress management when incorporated into daily activities, and as part of a wellness program, allows an employee to develop a strong sense of well-being both at work and at home because the employee learns to achieve a high degree of control over perceived stress.

It must be remembered that a great deal of perceived stress at work is a result of the cumulative effect of stressors which occur outside the workplace and whose residual effects are brought into the workplace by the employee. Therefore, an important component of any stress management program is educating the workplace population to recognize stress and identify its source.

Typically this is done in a small group setting with trained leadership to facilitate the sharing of information and the identification of stress inducing forces.

Once the stressors are identified the employees may then be shown a variety of methods by which these sources of stress may be successfully dealt with so that the perception of stress is reduced.

Except as noted above in regard to control, stress cannot be eliminated but it can be managed.

It is important at the onset of developing a workplace wellness program to evaluate the workplace population and determine the separate groups of employees that exist within the total population. Separate stress identification groups need to be formed which address stress as it pertains to their specific needs. For example, management employees will be faced with different work related stressors than would grounds maintenance employees.

It is best to contract with outside professionals when forming initial stress management groups. These individuals are normally found within the psychology and counseling fields. While the initial groups are best led by professionals, however, as employees successfully learn to deal with their stressors they may be trained to lead future groups thereby reducing the overall cost of the program.

It would also be wise to have a professional available for consultation even after co-worker leaders have been identified and trained.

Stress management is not a substitute for group therapy. Therefore, it is important that group leaders have the ability to recognize psychological symptoms which are obviously greater than those which may occur as the result of normal stress. When such symptoms are manifest, the employee should be referred via the human resources department to employee assistance resource professionals.

Stress perception evaluations may be completed during initial health screenings. The results of these evaluations will provide the necessary data for stress management group development, and will provide guidelines for the professional leadership in setting the format and goals of each individual group. Doing all evaluations at the onset of the program may be more cost effective than separating stress evaluations from general health screenings.

The psychology professionals contracted to lead the initial stress management groups can coordinate their evaluations with those being conducted by the health professionals involved in the initial employee screenings.

It is important for the wellness director to interview a variety of psychology professionals when initially designing the workplace wellness program. A variety of approaches are available to address stress. These approaches include communications analysis based programs, personality based programs (A versus B type personality), positive reinforcement training, biofeedback training, relaxation techniques, massage therapy, and so forth.

Prior to commencing the wellness program the wellness director should have a clear understanding of the locally available stress management options, and should have decided upon either one option in particular, or a combination of options which are most applicable to the employer's individual work setting and population.

When deciding upon options it should not be forgotten that, to a great extent, stress management is integrated within other aspects of the wellness program. For example: fitness and nutrition programs have very positive effects on normal stress reactions because the employee's ability to handle stress, without perceiving such stress as uncomfortable, is increased as a result of participation in the fitness and nutrition portion of the workplace wellness program. Therefore, specific stress management options should be chosen to supplement those which will exist as an integral part of the overall wellness program.

As with all aspects of the workplace wellness program it is important for upper management to show its support through direct participation in stress management groups. Adverse reaction to perceived stress is not unique to middle management and staff.

Employees at all levels generally feel that hard work is a positive way of showing that they are dedicated and loyal to their employer. However, when hard work becomes overwork, productivity and morale fall. Employees need to be trained to incorporate the principle of stress management into their daily routines. In this way they will be able to work more productively, and for longer periods of time, without becoming burnt out, overworked, or stressed out—all terms synonymous with perceived negative stress.

It is important for the employer to realize the benefit of assisting employees in the development of the ability to efficiently manage time, assess stressors, and separate themselves from their stressors in such a way that they do not take each stressor as a personal affront to their existence.

Once achieved, successful stress management is a primary key to individual and organizational success.

GROUP FORMATION

Individual group formation should not proceed until a stress evaluation survey has been done. This survey will provide data indicating which specific areas of perceived stress should be dealt with first.

Once identified, the areas of perceived stress should be evaluated from a pure loss control and production standpoint. Every effort should be made to deal with perceived stressors in an empirical manner before referring employees to a group for education in coping techniques.

Dealing with all perceived stress in a group format without due consideration of empirical solutions will project a condescending attitude toward employee concerns. If this occurs the group sessions will be ineffective, and will be ultimately counterproductive.

It must be remembered that stressors are both externally and internally generated. External stress within the employer's control must be dealt with first, then the employee can be trained to deal with internal stress reactions.

This is not to say that concomitant programs cannot be run. In fact, a dual approach may be ideal, but as noted above the groups should not be formed before environmental and production stress analysis has been completed.

In some instances such evaluations may uncover intrinsic operational problems which cannot be rectified by management for a variety of reasons, including budget, space constraints, local ordinances, and so forth. When this occurs, management should address these constraints with the workplace population in a candid and forthright fashion. Operational changes which are in fact feasible should be undertaken, and groups formed to help employees learn to cope successfully with those facets of the operation which cannot reasonably or practically be changed.

It is gratifying to realize that the workplace population will be willing to meet management halfway once it is understood that management is concerned about their well-being and committed to doing what can be done to mitigate identified adverse conditions.

Group formation will initially consist of employees who are dealing with individually identified stressors, with group constituency being made up of employees exhibiting similar stress related problems.

However, as time goes by the multiplicity of groups will gradually be reduced to one or two groups with dissimilar stress perception problems.

The reason for this change is that basically all types of stress are dealt with through similar educational techniques; however, during initial program formation there is a need to address individual problems and convince individual groups of employees that their particular problem is important.

Once stress management techniques are learned and used by the majority of the population, fewer groups will be necessary. As environmental and production problems are resolved, groups begin to deal more and more with internal reactions to stressors, and such internal resolutions are more generic in nature. In fact, at that point in the stress management program there is a benefit gained from nonhomogeneity of group constituency.

Once an individual realizes that it is not the stressor but their reaction to the stressor which needs to be dealt with, that individual is well on the way to successful coping. Nonhomogeneity contributes to this perception by showing that regardless of the diversity of the stressor the technique of coping is essentially the same.

This second phase of the stress management program then spreads through the workplace population and, once fully disseminated, results in employees who recognize the effect of stressors, are knowledgeable in coping techniques, and integrate those techniques into their daily lives.

Chapter 9

Prenatal Wellness, Childcare, and Elder Care

PRENATAL WELLNESS

Prenatal wellness activities for pregnant employees should be taken into consideration by the wellness director. It is a good idea to assess the interest and need for prenatal wellness activities during the initial and followup interest surveys which are conducted to determine program goals and program path development.

If it is determined from these surveys that prenatal care should be addressed, the wellness director can proceed with the development of a prenatal wellness program.

Healthy lifestyle habits which apply to the workplace population in general also apply to pregnant women. However, there are a few areas of concern which apply in particular to the pregnant employee.

Of primary importance is the fact that all pregnant women should be under the care of a physician. While this may seem obvious, it is unfortunately a fact that a great many women in this country fail to seek physician care until well into their second or third trimester of pregnancy.

This sometimes tragic circumstance is due either to the woman's lack of knowledge that she is in fact pregnant, or to a lack of education in regard to the importance of early physician care. In some cases, the pregnancy is unplanned and there is embarrassment and/or indecision in regard to whether or not the child will be carried to full term.

While the employer cannot be expected to act as the employee's guardian, it is reasonable to expect the employer to provide information to the pregnant employee about the importance of early physician care. Such information should also be included as part of a consent statement signed by the employee at the time she enrolls in the employer's wellness program.

A pregnant employee should not participate in fitness or weight management activities without her physician's approval. Specific guidelines which are to be followed by the individual pregnant employee regarding exercise and nutrition should be received from her treating physician. At no time should a program instructor or coordinator provide medical advice to participants.

Under normal circumstances a pregnant employee can participate in all regular wellness activities as long as she has conferred with her physician regarding possible contraindications. Under normal circumstances participation in wellness activities will be very beneficial to both the employee and her unborn child.

The wellness director should always make certain that a written doctor's release form is obtained and kept on file. The release form should specify which wellness activities the pregnant employee may participate in, and which activities should be avoided.

Exercise and nutrition are the areas of greatest concern in regard to the pregnant employee. Exercise during pregnancy continues to be controversial. There have been numerous studies done to determine the beneficial or adverse effects of exercise during pregnancy, and the results of these studies vary greatly. The consensus of the medical community at present is that exercise can safely be engaged in by pregnant women provided that its intensity is maintained at moderate levels.

Pregnant women who exercise tend to hold their pregnancy weight gain to within normal ranges, and they tend to return to their prepregnancy weight within a shorter period of time following child delivery than do nonexercising women.

However, it must be emphasized that moderation is very important. Excessive exercise can be detrimental to both the mother and the unborn child. Again the treating physician should provide appropriate guidelines for individual pregnant employees.

The American College of Obstetricians and Gynecologists have developed guidelines for exercise during and following pregnancy. The wellness program's fitness instructor should be familiar with or have access to these guidelines. Fitness instructors chosen by the wellness director should be well versed in addressing pregnancy concerns.

Chapter 9: Prenatal Wellness, Childcare, and Elder Care 69

It should be anticipated that very fit individuals who exercise regularly prior to becoming pregnant may have difficulty modifying their exercise routine appropriately when pregnant. Fitness instructors should be skilled in recognizing these situations and in dealing with them.

Ideally fitness classes designed specifically for pregnant women will address exercise modification concerns from an educational standpoint. Each pregnant participant should be counseled individually by the employer's wellness professionals to maximize the benefits they will obtain from participation, and to ensure that their physician's recommendations are being followed.

General guidelines also apply to pregnant women in regard to nutritional considerations. However, it is important to emphasize the need for individual physician approval of any special nutritional program.

Among knowledgeable health and fitness professionals it is commonly understood that dieting for weight reduction during pregnancy is not a good idea. Any exception to this general rule should be under specific guidelines determined by the pregnant employee's treating physician.

Dieting for weight loss during pregnancy or lactation is not recommended. The mother's body will return to normal body composition naturally with sound nutrition. This process should not be hurried by participation in dieting for weight loss, which might adversely affect the overall health of both the mother and her child.

Pregnancy is often a motivation for women to learn more about sound nutrition in order to provide their babies with the highest quality nutrients possible. The information about sound nutrition made available through the employer's wellness program will be a valuable adjunct both during and following pregnancy.

Lactation is also a very motivating time for the new mother to learn about nutritional principles. By applying the knowledge she gains from educational courses on nutrition, she will be able to provide herself and her baby the best possible nourishment. And she will find that her body's lean muscle to fat ratio will return to pre-pregnancy levels in a remarkably short period of time.

The wellness director and program coordinators can offer programs specific to prenatal concerns if the need and interest justifies the commitment. Including pregnant women in existing wellness activity programs can also be an appropriate means of addressing some of the pregnant employee's needs. However, it is important to utilize qualified personnel to develop and implement programs involving pregnant employees.

CHILD CARE

An increasing number of women are entering the work force with young children in need of daily care while their mothers are at work. Demographic projections indicate that this phenomenon will continue to increase through the end of this century and well into the next.

There is no question that productivity levels are adversely affected when a working mother is concerned about the quality of daily care that her child is receiving.

High quality child care is often expensive. Unfortunately, many working mothers find that they must sacrifice the standard of care they would prefer for their children in order to continue working.

In addition to reduced productivity there is an increased likelihood of an adverse stress reaction as the working mother finds herself coping with normal work related stress in addition to her doubts and concerns about the care of her children. There is consequently an increased probability that a workers' compensation claim may be submitted for work related stress. This probability is greatly increased if the employee is relatively low on the wage earning scale and comes to the conclusion that she could get by on workers' compensation benefits and be able to stay at home with her children. Consequently, many employers are taking a proactive stance in addressing these concerns.

One approach is to establish a child care facility at the workplace. Such a facility would cater primarily to pre-school age children, but may also be appropriate for early school age children as an after-school alternative to self supervision.

A few employers have combined senior citizen concerns with child care programs, partially staffing their facility with senior citizens who are parents and grandparents of employees with elder care concerns. This approach gives the elderly persons a chance to have the pleasure of young people in their lives and also fills up otherwise unproductive and lonely days.

These types of programs do not anticipate that the senior citizens will necessarily participate in the program on a daily basis, but rather schedule such participation in a way that it is looked forward to by all concerned rather than being seen as an unavoidable duty.

For the small to medium size employer an on-site child and/or elder care facility may not be an option. However, the wellness director can provide employees with assistance in finding outside resources. This assistance can range from a simple directory service to prequalification of providers and some financial assistance.

Often an employer or a group of employers can negotiate an employee discount at certain child care centers. The employer or employer group would be able to require specific levels of care quality and would represent a large enough financial block to assure that the providers would conform to the standards set.

Several employers might get together and help fund or staff a day care facility for children of their respective employees. Naturally, logistical and liability concerns must be considered, but these are not insurmountable.

Developing an activity program which can be participated in by the entire family will also prove quite beneficial.

ELDER CARE

Elder care is another area of concern for many employees. As with child care, these concerns often involve unexpected demands and emergencies. Often the concerns of elder care are more demanding and less easily resolved than those associated with child care. Further, the employee responsible for elder care who is no longer young may be ill equipped to cope with the energy demands of elder care responsibility.

Elder care is very stressful for many employees. Unlike child care there is little hope for the future. The demands of elder care will grow as time goes by until at last the cared for person passes on. This scenario can be very depressing to the employee, and can create a higher degree of stress reaction than that associated with child care.

Employers can assist employees through programs similar to those associated with child care. However, on-site facilities are normally not the option of choice.

Resource qualification and assistance is important, as well as enrollment of the employee in a stress coping program to aid in dealing with the pressures of elder care responsibility.

It should be remembered that an organization dealing with a variety of similar problems can more easily find resources and establish qualification criteria for care assistance than can one individual working alone.

Chapter 10

Hiring, Physical Examination, and Claims Management

HIRING

Traditionally, employee selection is not considered to be a part of a workplace wellness program. However, employees who are hired with preexisting medical conditions of which the employer is unaware may be placed into work situations which may aggravate the condition. A medical condition which is aggravated by work duties is a compensible claim under workers' compensation insurance law.

It is important that an employer make every effort to thoroughly screen all employee candidates. The preemployment screening process begins with the development of a thorough employment application. It is important for the employer to maintain a consistent policy in regard to the completion of written applications and the preemployment screening of employee candidates. No employee should be hired without having first completed a written application. The information provided by the employee candidate should then be thoroughly verified through a telephone and/or written verification process.

The verification of confidential information by the prospective employer falls under the jurisdiction of the Equal Employment Opportunities Commission, Labor Department, and/or other State and Federal agencies. Various regulations and rulings require an employer to disclose in writing to an employee candidate the employer's intention to conduct a verification check of the information provided on the application. Always include such a disclosure on any application on which verification checks will be performed.

Consistency of hiring policy is very critical. Once an employer has initiated a verification program it must be administered equally to all future applicants regardless of the nature of the employment being considered.

An employer who verifies applicant information on an inconsistent basis is running the risk of being accused of discriminatory practices. It is not the fact that verification is instituted that can cause problems for an employer; rather, the inconsistency with which verification is conducted that can lead to accusations of hiring discrimination.

For this reason, and to minimize possible exposure to other privacy violations, it is very important for the employer to establish a centralized verification procedure.

Many employers prefer to have line managers conduct the initial candidate interviews because these are the individuals who must direct the activities of the newly hired employee. There is nothing wrong with this approach; however, all candidates should fill out an application and the application should be forwarded to the person responsible for verification.

It is important that no representative of the employer offer a position or imply that a position will be offered during the initial interview.

All employee candidates should be informed that there is a three-stage process involved in hiring:

1. The initial interview
2. Verification of application information
3. Preemployment physical examination

If an employer's representative offers, or gives the impression of offering, a position to a candidate and subsequently the candidate is turned down as the result of poor verification results or failure to pass a medical examination, the employer may be placed in a position of being accused of discriminatory hiring practices.

The person chosen to verify employee candidate applications should ideally be:

1. Highly organized
2. A trusted employee, who has been with the company for at least one year

Chapter 10: Hiring, Physical Examination, and Claims Management

3. A person known for discretion and the ability to maintain confidentiality

Generally speaking an employer may verify any information given on the application by a prospective candidate, unless the candidate has specifically requested that a particular verification source not be contacted.

It is generally agreed that the following points may always be verified:

1. Employment dates
2. Duties
3. Salary
4. Eligibility for rehire

If eligibility for rehire is negative, it is a good idea to determine if it is a policy of the former employer, or only applicable to the candidate in question.

In speaking with a former employer, and in verifying dates of employment, it is prudent to verify that the next previous employer shown on the application agrees with the employment records of the verification source. This ensures the accuracy of the continuity disclosed on the application. Where there is an inconsistency a question arises concerning the candidate's reasons for lack of candidness in this regard.

Because each E.E.O.C. district administrator has a great deal of latitude in regard to the interpretation of discriminatory practices, it is very difficult to provide definite rules concerning verification questions which may or may not be asked. However, it is generally agreed that an employer may ask any question that relates directly to the work the employee candidate will be performing for the prospective employer.

Executive management should set confidential, nondiscriminatory parameters for the verification person to follow. The person verifying application information should be given the authority to rule on the advisability of hiring any applicant who does not meet predetermined hiring parameters.

However, if a line manager feels that an applicant is acceptable for a position, notwithstanding the decision made by the person in charge of application verification, then the decision to hire should be referred to an executive oversight committee for a decision. Both the verification person and the line manager should be given the opportunity to present their views in the matter. However, under no circumstances should the candidate be informed of derogatory information developed during a verification check.

PREEMPLOYMENT PHYSICAL EXAMINATIONS

Preemployment physical examinations are an essential aspect of developing a healthy workforce. The only effective means that a prospective employer has of determining the physical limitations of an employee candidate is through the utilization of a thorough preemployment physical examination.

Many employers perform perfunctory preemployment medical examinations. These are a waste of money. The examinations must address the specific physical qualifications necessary to perform the work for which the candidate is applying.

The preemployment medical examination should ideally be conducted by the same physician or group of physicians who will be participating in the wellness monitoring portion of the workplace wellness program. Ideally, these same physicians will also be called on in the event of an industrial injury or illness. The greater the continuity of medical appraisal, the more effective the results of the employer's wellness program.

In designing an appropriate preemployment physical examination the first step is an analysis of the work to be performed by the employee. Each job classification will have specific duties, and specific illness/injury exposures.

The employer should consult with the physician chosen to conduct the preemployment physical examinations and discuss the overall goals and philosophy of the wellness program, as well as provide the physician with job descriptions that include an analysis of the related exposures.

It is important that preemployment physical examinations include strength and endurance testing appropriate to the job classification exposure anticipated. It is equally important that an employee not be required to submit to such testing when such exposure is not anticipated. There have been cases of employers being successfully sued by employee candidates who were injured while being unnecessarily tested, or who claimed severe depression and mental anguish because they could not complete a test which was not related to the job duties for which they were being hired.

The employer should consult with the examining physician at the inception of the program and discuss the medical questions to be asked of the employee. These questions should include use of drugs, alcohol, or prior medically diagnosed conditions which might affect the ability of the employee candidate to perform the duties for which they are being considered.

The employer should not expect or desire the physician to disclose in detail the results of the physical examination prior to the employee being hired. Parameters of qualification should be established ahead of time, with the physician placing the candidate in predetermined categories relative to the candidate's ability to qualify medically for the position being offered. Such parameters must be applied consistently and in a nondiscriminatory fashion.

In this way the employer is not placed in a position of learning medical facts which are a matter of privacy. And the physician is not placed in the position of making the hiring decision for the employer.

Testing for the use of controlled substances should also be performed at the time of the preemployment physical examination. Again, full disclosure and candidate agreement is necessary at the time the employment application is completed.

Controlled substances testing must include the following laboratory controls: The bodily fluids sample should be tested by a laboratory certified by N.I.D.A. to perform such tests for a variety of employers. Always talk with other employers to verify that the testing laboratory has performed satisfactorily for them. Always get client references from the laboratory.

If the employee candidate requests the results of the test, they should be provided with such results. The candidate may protest the results on the basis that he or she was or is taking a prescription drug. However, this information should have been developed on the preemployment physical examination questionnaire. In any case, the candidate's contention of prescription drug use should be confirmed, both in regard to its validity, and in regard to the prescribed drug's ability to affect the results of the controlled substance testing.

In order for an employer to refuse to hire an employee on the basis of a positive controlled substance test, the employer must have a written policy of zero controlled substance use tolerance. Further, the policy must state that positive test results are a cause for nonhire or termination. Employers are encouraged to provide candidates who have tested positive with referrals to local private and/or community substance abuse rehabilitation programs.

Preemployment physical examinations need only be performed on employee candidates who have successfully completed the first two stages of the employment process as noted above.

WORKERS' COMPENSATION CLAIMS MONITORING

It is not within the scope of this work to cover all of the loss control aspects of workers' compensation claims management. However, we do feel that a general discussion is in order.

Workers' compensation claims can drastically affect the profit margins of an employer. Control of claims through claims prevention and claims management is extremely important.

The first step in managing workers' compensation claims is the establishment of a designated physician program.

Workers' compensation law in many states allows an employer to choose the physician who will treat an employee in the event of a work related illness or injury. The employer may require the employee to continue treatment with the designated physician for a period of 30 days following the date of injury. After that time the employee may choose treatment with another physician if unsatisfied with the initial treatment.

Workers' compensation law in some states also allows an employee to choose his or her own physician for treatment of a work related illness or injury if the choice is made before the injury occurs.

It is best to resolve the issue at the time of employment by asking the employee to agree to see the employer's designated physician in case of a work related illness or injury. If the employee at that time states that he or she has a family physician then this should be noted.

(Since workers' compensation law varies from state to state, it would be prudent for the employer to consult with appropriate legal counsel prior to establishing the designated physician program.)

In general, employees will be willing to agree to be seen by the employer's designated physician if it is explained that:

1. The physician specializes in industrial illness and injury and is familiar with the employer's operations
2. The employer is concerned about the employee receiving the most appropriate treatment as quickly as possible if they are injured while working
3. The employee who is subsequently unhappy with the treatment received may choose another physician

When this is properly presented, virtually all new employees will agree to being treated by the designated physician.

Although workers' compensation law allows a physician to disclose medical information to an employer relative to a work related illness or injury case, it does not require such disclosure. A physician designated by the company is more likely to provide candid assessments of an employee's progress than is a physician who may feel his or her primary loyalty is to the patient, notwithstanding the nature of the illness or injury.

TREATMENT FOLLOWUP

Treatment followup is a very important aspect of workers' compensation claims management. An employer should bring an employee back to modified duty as soon as possible after a work related illness or injury has occurred. The sooner the employee returns to work the less the cost that will be incurred.

Early return to work programs are greatly reduced in their effectiveness without the complete cooperation and coordination of the treating physician.

EARLY RETURN TO WORK PROGRAMS

Early return to work programs must be coordinated with the employee's treating physician and the employer's workers' compensation insurance carrier. The assignment of modified duties is not a unilateral decision to be made by the employer. Even where work related injuries are relatively minor the employer should have a written doctor's release for the employee to return to work; the release should state the limitations of the work duties the employee is allowed to perform.

Never allow an employee, other than in cases of minor first aid treatment, to return to work without a written doctor's release. Without a release which indicates restrictions, or no restrictions, the employer runs the risk of incurring liability by assigning work duties which exceed the injured employee's limitations.

Some employees will enthusiastically return to work, overestimating their own ability to work following an injury. Later, if the injury is permanently aggravated, they may claim that they were pressured into such an overestimation.

When a person sustains an injury, the first reaction may be one of embarrassment and self-blame. Later, especially if there is a substantial period of recovery required, the individual will feel anger, often directed toward the employer.

Additionally, regardless of the injured person's long term reaction, he or she will have distinct feelings of failure and inadequacy. If the injury causes a degree of permanent disability, the injured person may feel unable to again make a satisfactory living, may feel alone and isolated from former co-workers and friends. For many people, a job is more than just a means of obtaining funds for living. Work duties may provide significant internal satisfaction, and the daily routine may provide important social benefits. Depression is not at all unusual after an injury which results in lost work time.

The employer's treatment followup procedures, early return to work program, and wellness program can in many cases completely alleviate these post trauma symptoms. A telephone call to see how an injured worker is doing and/or a get well card from associates can make all the difference in the world. Of course, if these overtures are perfunctory and uncoordinated they will have little long term effect, but when they are coupled with a well organized program of concern and rehabilitation the results can be quite spectacular, and rewarding in a way that is not measurable in dollars alone.

The goal of the early return to work program is to fully rehabilitate the worker to his or her former position of productivity and responsibility. In some cases this is not possible. Then the employer must determine if the employee is qualified for other work within the company, or must be trained to perform other duties outside the company. Again, coordination with the employer's insurance carrier is paramount.

Finally, there are some accidents from which there can be no recovery, and there are some employees for whom there can be no rehabilitation because of the nature of the injury, or the capacity of the individual. This is the toughest situation for the caring employer to have to face. Again with input from both the treating physician and the insurance carrier's return to work specialists these unfortunate circumstances are mitigated to the extent that the employer does not need to carry the burden of the decision alone.

Chapter 11

Insurance Options

One of the benefits to be derived from an effective workplace wellness program in conjunction with an effective workers' compensation loss control program is a reduction in overall insurance costs.

To understand how the above programs indirectly and directly reduce insurance costs, it is necessary to understand the basic insurance options available to an employer.

Employers are divided into two groups: those who are self insured and those who are not. A self-insured employer must meet specific requirements as set forth by the state in which they are incorporated, or in which they maintain their home offices if they are not incorporated.

Although the requirements vary from state to state, generally the employer will be required to post a bond, or establish an insurance reserve to qualify as a self-insured entity. For worker's compensation self-insurance one million dollars capital reserves are generally required.

Additionally, the employer must prove the capability to manage a self-insured program, or must retain the services of an insurance management firm that meets state requirements for insurance management expertise.

A captive insurance company is a corporation formed to manage the investment of premium dollars ceded to the captive for insurance reserve purposes. The term *captive* refers to the fact that although the company may be chartered as an insurance company it has only one insured, the employer, or a group of similar employers who originally formed and invested in the captive.

Off-shore captive chartering at one time offered distinct tax advantages in regard to investment profits which have mostly been eroded through changes in applicable Internal Revenue Service regulations. In discussions of captive insurance companies, the term *captive* is often equated to an off-shore entity, which is not correct. A captive may be formed domestically or off shore.

A captive insurance company is formed by employers who are both the founders and also the insureds. A "rent a captive," however, may be formed by a group of investors in order to offer insurance to a limited and narrowly defined group of insureds through an insurance agency, bank, or other financial or insurance management firm.

One common link between a true captive and a "rent a captive" is the homogeneity of the insured exposures. That is to say that only certain types of employer risks are allowed into the captive, depending upon underwriting prerequisites established at the time of the captive's formation. A standard insurance company, in contrast, may underwrite a wide variety of employer risks.

Employers considering captive options in the hope of having greater control over claims management, reserves, investment, and dividend distribution should explore all options with extreme care. Control of funds by "rent a captive" management may be no better than would be found in a group insurance option through a standard insurance company, and only slightly better than a standard insurance policy.

A similar caveat would hold when forming a captive in conjunction with a group of other employers. The employer with the largest premium ultimately exerts the greatest control.

We will begin our discussion with those insurance options available to the small employer, as defined above, and then progress to those options available to the large employer under a self-insurance program.

For the small employer, strategies for reducing insurance costs revolve around the need to obtain insurance through established insurance providers at the lowest possible premium.

In all cases the first step in reducing insurance costs is a reduction in the frequency and severity of claims reported. For the small employer the most easily recognized direct benefit of a reduction in claims frequency and severity will be seen in the employer's workers' compensation experience modifier.

In all states the insurance premium charged for workers' compensation insurance is based upon rates established by the state insurance department. These rates are established after statistical analysis of all worker injuries which have occurred within the state's established risk classifications over a specified period of time. This rate is called the *manual rate,* referring to the state's workers' compensation insurance rate manual. The rates are normally set for each one hundred dollars of payroll.

A state with injury statistics for a given risk classification which is lower than the national average will normally have lower manual rates for the given classification than another state with injury statistics which are higher than the national average.

The actual rate that the individual employer is charged is then modified to reflect the individual employer's workplace injury record. For example, say an employer's premium, based upon manual rates for the risk classifications of their employees, equals $100,000. If the employer has a good claims record, they might receive an experience modification of .75. The employer would, therefore, pay a Workers' Compensation Insurance premium of $75,000 instead of the manual premium of $100,000 — a substantial savings of $25,000.

However, an employer with a poor claims record might receive an experience modifier of 1.75. This employer would, therefore, pay a Workers' Compensation Insurance premium of $175,000.

The insidious point which must be remembered is that the employer paying $175,000 in premium is not just paying $75,000 more than the premium set by the state manual. The employer is, in fact, paying $100,000 more than would be required with a good claims record.

It should be noted that in most cases the cost of funding a combined workers' compensation loss control program and a workplace wellness program is far below the cost of increased workers' compensation premiums as the result of increased experience modifiers.

Many work related injury claims are the result of employees not maintaining an optimal personal health profile.

Workplace wellness programs enhance safety and loss control efforts which directly affect the work related injury and illness rate. However, even the best loss control program may not reduce claims which in fact begin outside the workplace. A workplace wellness program on the other hand will address many of those claims, thereby further reducing the total claims associated with workers' compensation insurance.

The premium quoted for personal health benefit policies is also modified based upon claims charged against the policy. However, unlike workers' compensation modifications, these premiums and modifier formulas are not generally mandated by state insurance departments.

Employers with established workplace wellness programs will have lower claims frequency, and may therefore be able to negotiate more competitive rates. Some employee benefits insurance companies may in fact offer rate discounts up front to employers with effective wellness programs in place, and this is an important feature to consider when shopping for insurance.

After the employer has taken the necessary steps to reduce claims frequency, the next step in reducing insurance costs is the careful consideration of insuring options.

Sources of insurance fall primarily into two categories:

1. Independent insurance agents and brokers
2. Direct writing insurance companies, and insurance company agents

Again let us begin by defining terms. An independent insurance agency is a business which offers insurance policies underwritten by a variety of insurance companies. The independent agency is not required to offer insurance packages from only one insurance company, nor is it obligated to give first choice of refusal to any one insurance company.

The difference between an independent agent, and an independent broker is as follows: an agent is appointed by an insurance company to represent its products. The agent, in return, is often given greater underwriting consideration than an independent broker. However, this is not necessarily the case. Often the only difference between an independent agent and an independent broker is the commission structure agreed upon between the insurance company and the agent or broker.

An independent broker does not have an insurance company appointment to act as its agent. This may not necessarily be a drawback unless an employer's insurance is difficult to place because of loss exposure. Normally, an independent agent with an appointment to an insurance company will be given primary consideration over a broker without an agency appointment. Therefore, if your business is difficult to write and the insurance company must choose between the same exposure difficulty being presented by one of their appointed agents as opposed to a nonappointed broker, the agent will normally receive first consideration and possibly a more competitive quote.

On the down side, agents are required to meet certain preestablished written premium quotas in order to qualify for certain levels of commission. Hence an agency might bypass a nonappointed option that a broker would otherwise select.

Practically speaking, independent agents are also independent brokers. As an insurance purchaser the employer should make the point of discussing all available options with an agent/broker including those insurance companies with whom the agent has appointments and those other insurance companies who may also service the employer's area of risk with whom the agent does not have specific agency appointments.

The employer, in conjunction with the agent, should be aware of which insurance companies will receive an application submission, and then, at the time of quote, should discuss the results of those submissions, specifically, which companies refused to quote because of underwriting concerns, as well as the options of those insurance companies that did present a quote.

The employer should never allow an agent to present only one quote, unless all other insurance companies discussed at the time of application submission declined to quote on the employer's business.

While an employer may consider it wise to approach more than one independent agent when seeking insurance quotes, it is very important to ensure that all agents approached receive and submit identical application information. Always ask to see a copy of the agent's application at some point during the submission/quote process to ensure the accuracy of the information submitted. A few unscrupulous agents/brokers will withhold application information from insurance company underwriters to improve their chances of receiving a more competitive quote. Ultimately this hurts the employer because inevitably the insurance company learns of the discrepancy, and will then assume that the employer has withheld the information from their agent and will adjust their future renewal quotes, and underwriting thinking accordingly, or will cancel the policy flat.

By the same token there is absolutely no value in an employer withholding information from an agent/broker. It will be discovered, and will ultimately result in increased expense to the employer, and/or difficulty in obtaining future insurance coverage at reasonable cost.

When seeking multiple quotes from multiple agent/brokers, always find out in advance which insurance companies will be receiving applications. There is no value gained in allowing two different agent/brokers to submit applications to the same company. This only creates confusion. There is no value gained in dealing with more than one agent/broker appointed to, or submitting applications to the same insurance company. The result is that one agent blocks the other agent's market, delaying or eliminating entirely the quote sought.

Always find out which markets the agent intends to submit an application to, then choose the markets which you as an employer would like the agent to pursue.

Never disclose to any agent the names of other agents with whom you are also seeking quotes. Some agents will automatically submit applications to their competitor's market in an often successful attempt to block the other agent's market. This ultimately reduces the employer's range of quote options.

Another source of insurance quotes is the exclusive agent system. Some companies have exclusive agency appointment agreements which require their agents to allow them first right of refusal. These agencies are easily identified because the agency name will almost always incorporate the name of the insurance company holding the exclusive agency agreement. These agencies may also broker business to any other market once their primary market has refused to quote on the business, or if their primary market's reinsurance treaty does not allow it to underwrite the risk classification presented by the employer's business.

However, exclusive agents, by definition, do not have multiple agency appointments, therefore they are strictly brokers outside their exclusive appointments and the benefits and drawbacks of brokers versus agents as noted above would apply.

Insurance coverage may also be obtained directly from some direct writing insurance carriers. Frequently, however, direct writing carriers will only consider policies which will develop premiums in excess of $100,000.

Many insurance companies will provide coverage through both independent agents and direct writing facilities. If, as an employer, you feel that your company might qualify for direct writing coverage then this option should be pursued prior to submitting applications through independent agents. An insurance company which writes directly as well as through independent agents will normally not quote on an employer's business directly if an application has already been submitted through an independent agent.

Obviously, insurance options are business decisions which require long lead times. An absolute minimum lead time for submission of applications either through an agent/broker or directly is 90 days. A wise employer will begin the search process 12 to 24 months in advance, investigating the options by choosing the agents through which to submit applications and by exploring direct writing opportunities as well.

Ideally, 120 days prior to renewal the application process should begin through all identified sources. It should be anticipated that quotes from the insurance carriers will not be available sooner than 10 days prior to renewal. All insurance carriers realize that a wise buyer will be carefully evaluating their options. Carriers are careful to avoid getting trapped in a quote war by narrowing the negotiating period down to the smallest number of days possible. Unfortunately, neither the employer nor the agent has any control over the carrier's perogative in this regard.

The cost of the insurance policy sought is not necessarily the most important factor, and may ultimately not be the determining factor for an employer in choosing where to place their insurance business.

The first assessment the employer must make is in regard to their own in-house capability of handling claims and other communications problems that will arise between the insurance carrier and the employer. Unless the employer has employees knowledgeable in dealing with matters of insurance, and has allocated the time and budget necessary for the resolution of these issues, the option of choice should always be in favor of placing the insurance coverage through an independent or exclusive agent.

In choosing agencies it is important that the agency have the capability to service the employer's needs adequately. Depending upon the complexity of the employer's insurance program the agency chosen will need a variety of services.

Full time customer service representatives should be on duty at the agency during normal business hours. They will be capable of handling the communications problems that arise between the insured employer and the insurance carrier. Small agencies which employ multiple-duty personnel are in no position to service the needs of the medium to large employer.

The agency should have a claims person or department. In this instance, customer service representatives will often fill the roll of claims representatives as well, and this is normally acceptable. However, agencies with CSRs who must also act as office managers, receptionists, and secretaries, are not acceptable, as their time and energy will inevitably be spread too thin to service the employer's account adequately.

In general, the greater the variety of services offered, the more professional the agency, and the better the service the employer will receive. An employer who chooses price exclusively over agency service will ultimately incur greater costs as insurance related problems arise.

Employers should realize that there are no standard insurance contracts. While there are great similarities between certain types of coverage and insurance companies, there are also significant differences.

The differences are highlighted by the exclusions stated in each individual insurance policy or reinsurance treaty. Most buyers make the mistake of comparing coverages and premiums while ignoring exclusions. This is understandable because many insurance agents fail to read and understand the exclusions stated in the insurance policies they are selling. Therefore the employer must ask what the exclusions are, and if deciding between two or more policies how the exclusions compare in each policy. Often the more expensive policy is more expensive because it has fewer exclusions.

Since there are no true standard insurance policies, the employer should make every effort to customize their policy to their individual needs by negotiating exclusions and deductibles either through their agents or on their own behalf if purchasing the policy directly from the insurance company.

Underwriting managers generally have the authority to customize certain aspects of their company's policies by adding endorsements which cover those situations not covered or excluded in the company's standard policy.

Again it should be emphasized that such negotiations coupled with sophisticated risk management and custom tailored deductibles can substantially reduce the overall cost of the employer's insurance program.

THE SELF-INSURED OPTION

Employers who qualify as self-insured entities must either meet their state's requirements in regard to insurance management expertise, or retain the services of outside firms.

In-house insurance management is by far the preferable choice, because it ensures that the employer will have ultimate control over the management and disbursement of funds reserved for the self-insurance program. Therefore, any employer electing to become self-insured should make every effort to meet the requirements of their state in regard to insurance management expertise by hiring individuals with the requisite qualifications.

Although true self-insurance may be outside the reach of many employers, a quasi-self-insured program is quite feasible in many instances where employee benefits are concerned.

The premium charged for employee benefits insurance is greatly affected by the type of coverage chosen and the level of deductible selected. That is to say that the premium for a major medical policy with a $250 deductible will be higher than that charged for a major medical policy with a $1,000 deductible. By investigating the deductible options, an employer may find that self-insuring a portion of the employee's deductible can greatly reduce the overall premium cost while still allowing the employer to remain competitive within the employee benefits marketplace.

This approach coupled with a sound risk management assessment of which exposures need coverage, extent of needed coverage, and which exposures can be self-insured (left uninsured) will offer the employer the most options in reducing the overall cost of their insurance package.

Risk management is a fluid assessment which changes from year to year. Generally, risk management departments work closely with loss control and wellness personnel to adjust their insurance decisions according to the effectiveness of loss control and wellness programs as well as the changing nature of the employer's business exposures.

Typically a growing employer, or an employer who is just beginning to develop a comprehensive insurance cost control program, will transition through the following phases:

1. Selection of an insurance agency for maximum service and best price option; a full service agency may also be able to provide loss control assistance; such assistance would be an intermediate step prior to the hiring of a risk manager
2. Development of a loss control program
3. Development of a workplace wellness program
4. Development of a risk management department
5. Establishment of a quasi-self-insured program
6. Selection of an outside self-insured management firm, or the development of an in-house insurance department
7. Transition to full self-insured status

Certain coverages, such as errors and omissions and product liability, are best underwritten through insurance carriers because of the unpredictable nature of punitive awards. These coverages generally have high deductibles and therefore take on the nature of quasi-self-insurance coverage from a practical standpoint.

Chapter 12

Ergonomics

Egonomics is defined by Webster's dictionary as "the science that seeks to adapt work or working conditions to suit the workers."

Twenty years ago the application of such a science in the workplace was unacceptable to most employers. Today it is the catch phrase of the decade. In the foreseeable future, ergonomic design considerations will be a normal part of virtually all employer operations.

This enormous change in attitude has come about primarily because of the escalating costs and high frequency of cumulative trauma claims.

When the worker's compensation system was originally conceived it was designed as a system to treat and compensate workers for acute traumatic injury. However, as the years went by this initial concept was expanded to include illnesses which were related to workplace environmental exposure, and/or manufacturing process substance exposure.

Because many illnesses are by definition cumulative, there occurred an overlapping of the boundary between illness and trauma. The first cases to develop out of this confluence of illness and trauma were spinal column claims. Submission of spinal claims rapidly expanded to include the muscles and tendons of the upper and lower back.

Once the definition of compensable trauma evolved past acute injury and into cumulative injury, the door was opened for broader definitions of work related illness and injury.

Consequently, disabilities which in the past were considered strictly a result of natural aging processes are today routinely accepted as compensable claims within the workers' compensation insurance system. Bureau of Labor Statistics data available at the time of this writing indicate that 50% of reported work related injuries involve cumulative trauma or illness disorders.

It follows that if a significant portion of the cause of claims frequency escalation is a de facto redefinition of normal aging, then the solution is to design a workplace which reduces the long term cumulative trauma experienced by workers and hence extends the amount of injury-free work time available to aging biosystems.

Fortunately for the employer, investment in ergonomic changes produces concomitant production efficiencies which in turn increase profits. Further, under ideal circumstances ergonomic changes will reduce the hiring turnover of skilled workers, resulting in a more experienced and steady workforce and a significant reduction in new worker training costs.

While the design of ergonomic workplace systems is more directly related to the field of loss control, it will be briefly discussed here in regard to certain specific tasks which overlap between the workplace environment and the nonworkplace environment.

Since workplace wellness programs are designed to promote increased worker health and fitness in order to provide a stronger and more fit workforce, it follows that allowing a worker to subsequently work in an environment which is conducive to cumulative trauma is diametrically opposed to the goal of workplace wellness.

It is beyond the scope of this work to address every possible workplace condition which can and should be addressed through ergonomic design. However, we would like to address two specific areas: computer workstation ergonomics, and the reduction of probably injury to a worker's back through training and ergonomic change.

In the second category especially, there is a strong overlapping of cause between work related injury and non-work related injury. It is in this category that we see the most frequent misuse of the workers' compensation system.

We also see extensive use of employee health benefit systems due to back injury occurrences outside the workplace. As was noted earlier, the employer's costs are increased regardless of where such an injury is placed. Ideally, avoiding the injury altogether is preferable to placement in either benefit system.

COMPUTER WORKSTATIONS

Cumulative trauma associated with computer workstations is currently one of the fastest growing areas of concern in the workers' compensation system. In fact, it has reached a point of such visibility that a number of municipalities have passed ordinances requiring employers to make appropriate ergonomic changes in the computer workstations used by their employees.

Attendant publicity has created an atmosphere which ensures that there are virtually no clerical employees within the United States who are not aware of the cumulative trauma disorders attributed to non-ergonomically designed computer workstations.

Any employer who has not currently addressed these concerns within the workplace may expect to experience an associated cumulative trauma claim within the foreseeable future.

Corrective action begins with design of the space designated for computer workstation use, and with the furniture associated with use of that space. Both the space and the furniture should be designed to reduce employee fatigue and discomfort. It should promote comfort, and a feeling of well being, which will in turn lead to a high level of worker productivity.

Long periods of time in the posture of sitting, particularly where the option of standing and moving about is limited, leads to rapid fatigue and discomfort.

Studies by Dr. Marvin Dainoff of the Ergonomic Research Center indicate that office workers who work for long periods of time at poorly designed computer work stations are at significant risk of incurring cumulative muscular disorders, particularly in the areas of the lower back, the neck, and the shoulders. In addition, depending upon many factors including genetic proclivity, acquired cumulative disorders of the wrist appear to be associated with long term use of keyboards in poorly designed workstations.

Coupled with the problem of wrist disorders are many other disorders which may have only vague symptoms initially. These symptoms include: higher than normal fatigue, headaches, tingling in the upper extremities, and localized pain which becomes more pronounced and severe as time elapses.

At first the worker attempts to ignore the symptoms; later productivity becomes adversely effected, and eventually the worker ceases working for the employer and/or submits a workers' compensation claim.

Fortunately, in most instances, these kinds of disorders are easily avoided by training workers to sit at their work stations with correct posture, and further by the use of appropriately designed furniture and workstations.

While each workstation situation should be individually analyzed by a person qualified in ergonomic job analysis, the following guidelines are generally applicable.

1. When the worker is seated with correct posture the video display terminal should be at eye level. A screen which is too low requires the user to bend the head downward. This places significant stress on neck and shoulder muscles, which rapidly leads to fatigue and subsequent pain.

2. Forearms should rest comfortably on a padded rest with the elbows bent at a 90 degree angle; wrists should always remain in line with the forearms. Unsupported forearms exert continuous downward weight on shoulder muscles, leading to rapid fatigue. The problem is compounded if shoulder muscles are required to either lift the shoulders, or lower the shoulders in order to access the keyboard.

Workers should avoid resting the palms of their hands on work surfaces in such a way that the wrist is bent. They should take frequent work breaks by standing up at the workstation and performing a variety of stretching exercises. The appropriate stretching exercises can be taught within the context of a workplace fitness program.

At the time of this writing, computer accessory manufacturers are beginning to market forearm rests which are designed to fit virtually any workstation.

3. Knees and hips should remain parallel to each other to evenly distribute upper body weight on the spinal column and buttocks. Many workers have the habit of curling one leg under their buttocks and sitting on it. Such a seated position can result in a variety of cumulative disorders to the leg, foot joints, and supporting muscles.

4. Feet should always be supported, either by a foot rest or directly by the floor. They should never be allowed to dangle. The result of dangling feet is restricted blood flow to the lower extremities, and subsequent nerve and muscle damage due to cumulative blood circulation problems.

Footrests are a very valuable but inexpensive accessory. Frequently, even when the very best attempt is made to purchase a chair which can be adjusted to any individual's needs, it will be found that those people who are shorter than average cannot create an adjustment compromise between work surface heights, chair seat height, and distance to the floor. In this situation the footrest is an ideal solution.

Footrests are also very valuable for average to above average size people because they allow some variety in leg angle which in itself reduces fatigue and improves circulation. A footrest should be available at all work stations.

Furniture

The most important piece of furniture associated with keyboard and video display screen usage is the chair. A chair directly supports, and when properly designed aids in the maintenance of good posture. The choice of any chair should be based on the size of the individual who will be using the chair, and the task for which its use will be required. An appropriate chair is not a perquisite, and therefore job titles should have no bearing on chair selection.

Any chair chosen should have the capability of being easily adjusted by the employee while seated. It should be possible to tilt the angle of the chair back and the angle of the seat independently. The back of the chair should be large enough to fully support the user's back, and preferably adjustable to a variety of back configurations. In particular, lumbar support should be individually adjustable. However, the back of the chair should not be so large as to restrict the movement of the arms.

The seat should be wide enough to comfortably support the user, and should be slightly concave in order to evenly distribute the user's weight. The leading edge of the seat should be curved downward to reduce pressure on the backs of the legs and improve circulation while seated. The seat's padding must be soft enough to allow some adjustment to body contour, yet be firm enough for adequate support.

It is important to make certain that the highest quality material is being used in the construction of the seat. This hidden factor is often overlooked by purchasers, and is an area where manufacturers can reduce costs while maximizing profit margins. Make certain that seat composition is thoroughly explored and understood before a purchasing decision is made.

Arm rests should be optional at the discretion of the user. When keyboard forearm rest accessories are used in association with the work station, chair arm rests may in fact be inappropriate. Chair arm rests tend to strike work surfaces when users attempt to rotate the chair seat while seated in close proximity to the work surface. This problem can lead to unnecessary marring of furniture, jarring of electrical equipment placed on the work surface, and under a worst case scenario lead to a low back injury due to the sudden twisting motion imposed upon the user when the seat stops short while the user's upper body continues to rotate after the chair arm strikes the edge of the work surface.

If keyboard forearm rests are inappropriate for a given workstation's configuration, then chair arm rests should be provided.

If chair arm rests are chosen they should be height adjustable, padded, and also have the capability to be swung out of the way, or easily removed.

Work surfaces, especially those associated with secretarial desks with attached typing returns have varied surface heights. It may be necessary for the user to adjust the chair height frequently depending upon which surface is being used. If the chair height is not easily adjustable while seated, the user will not adjust it as frequently as needed, but will rather attempt to adapt body height through muscular adjustment. This will ultimately lead to cumulative injury. Chair users should be trained and encouraged to adjust chair height as frequently as needed.

Chair bases should consist of five caster support arms to assure stability and reduce the possibility of tipping to a minimum. The casters chosen should be designed to roll over carpeted or uncarpeted surfaces with equal ease.

Fabric upholstery which is woven in such a way that it is not too smooth should be chosen. This type of upholstery reduces temperature buildup between the user and the seat surface and avoids the need for users to tense their muscles in order to avoid slipping along the surface of the chair seat. Plastic or leather upholstery is not desirable from an ergonomic point of view.

The upright human body is by nature unstable. It therefore requires continuous muscle activity to maintain an upright position.

Gravitational forces exerted on the lumbar vertebrae and disks are greatest in the seated position.

An ergonomically designed chair supports the seated person in such a way as to reduce to a minimum the gravitational stress imposed upon the neck, shoulder, back, and leg muscles.

Document holders are important to reduce neck and shoulder strain. Work materials placed in document holders should be at the same level and angle as the video display screen. The need to refocus between video display screens and hardcopy material slows work performance. This slowdown is most dramatic in older workers, whose eyes naturally take a longer period of time to adjust, than in younger workers. However, in all workers the need to refocus slows productivity and increases the likelihood of eye strain related headaches.

Task lighting is extremely important. Single source overhead lighting is not recommended in video display work situations. The use of fluorescent lighting is in particular not recommended in conjunction with video display work stations for two reasons: (1) as with any single-source overhead lighting, lighting which is bright enough to illuminate work surfaces adequately will generate unacceptable glare on the video display screen, often even when antiglare screening is used. (2) The nature of fluorescent lighting is such that as the light source ages a flicker is generated. The flicker is normally well above the ability of the human eye to recognize; however, when the flicker is coupled with the scan rate of the cathode ray tube which "paints" the information displayed on the video display screen, user discomfort may be experienced.

Video display workstations should be established in areas of subdued overhead lighting, coupled with subdued workstation color decor designed to evoke calmness, not to be confused with relaxation.

Supplemental work surface lighting which is shaded to eliminate video display glare should be provided. the least stressful supplemental light is generated by sources other than fluorescent sources, notwithstanding the energy savings achieved by fluorescent lighting.

If natural lighting is used to illuminate video display work stations the screen should be turned 90 degrees to the source of light. The user should not face directly toward the light source, nor directly away from the light source. Video display glare screens must be used if natural light sources are used in order to reduce screen glare which will occur at some point during the day, as the ambient light changes due to the earth's rotation or the waxing and waning of cloud cover.

A great deal of publicity has recently been generated concerning the health risks of low frequency radiation generated by electrical appliances, personal computer systems, and video display screens. At the time of this writing no definitive studies have confirmed the existence or extent of such health risks. Therefore, we do not address this concern here. However, we do not claim that such health risks do not exist, nor do we state that at sometime in the future empirical proof will not be found.

It is suggested that prudent employers should take such precautions as are currently available and reasonably possible within the context of their operations, and should maintain an open attitude toward the possibility of at least a subtle risk existing from such exposure.

Keyboard rests and mounts should be adjustable both in height and angle of access.

Work surfaces should also be adjustable, but not necessarily by the user. The work surface should be adjusted to the user's height, based upon manufacturer's recommendations, at the time the user is assigned to the workstation. As much as possible within the context of an employer's operation, sharing of workstations should be kept to an absolute minimum. When sharing is necessary, shared workstation assignment should attempt to place users of similar height at shared work surfaces.

Each time a new employee is assigned to a work surface with the intent of long term assignment, the work surface should be adjusted to the employee's height even if a prior user was nearly the same height.

Summary: Video Display Work Stations

- Elbows rest at 90 degree angles when addressing keyboards.
- Legs, ankles, and knees should be parallel to each other when seated.
- Neck, back, and waist should be in line and as straight as possible, receiving appropriate support from ergonomically designed furniture and equipment.

- Feet should rest flat on floor, or on a footrest.
- Shoulders should be kept level and parallel to the floor's surface.
- Make certain workstations are designed to accommodate the tasks which will be carried out within each station. Do not attempt to increase the workstation task usage without taking into consideration ergonomic principles.
- Work surfaces should be adjustable to height of user.
- Provide foot and padded arm rests.
- Adjust document holders to video display screen height and viewing distance.
- Tasks should always be performed close to the body and require light movements.
- Ensure video display screens have brightness and contrast controls, with antiglare screens where appropriate.
- Chair height must be adjustable by user while seated. Train and encourage frequent adjustment.
- Keyboard height and angle of access must be adjustable.
- Task lighting should be appropriate. Single source overhead lighting is not recommended.
- Appropriate orientation of video display screen to lighting source(s) is necessary.
- Appropriate supplemental work surface lighting is necessary.
- Set productivity quotas appropriately. Avoid stress from inappropriate pacing. Seek input from employees.
- Rotate work tasks to avoid cumulative trauma from continuous, long term, repetitive tasking.
- Provide frequent short work breaks which allow employees to stretch their muscles at their work stations. Train and encourage employees to utilize these break times appropriately.
- Encourage responsibility in each employee. Seek employee input on pacing and work station design. When pacing, work station design, furniture, and equipment are chosen, educate employees in regard to the reasons decisions were made. Emphasize the employer's concern for the health and well-being of the worker. Such education should be a part of all new employee orientation.
- Keep an open attitude about ergonomic design changes. The field of ergonomics is constantly changing. Adapt new principles when and where feasible.

BACK CARE

Injuries involving the spinal column, or strain and sprain of the muscles, tendons, and ligaments of the back, are perhaps the most common and certainly among the most costly injuries sustained by workers.

Many workers sustain back injuries over a period of time as the result of poor lifting technique both at work and at home, or while participating in recreational activities.

It is important that all workers receive training in proper lifting technique, as well as education in basic back physiology and the causes of back injury. Many employers are surprised at the number of clerical workers who sustain a back injury in spite of the fact that lifting is not part of their job description. Clerical employees, in fact, do a fair amount of bending and lifting, including accessing reams of copier paper, which may require moving a case-size box weighing up to 50 pounds, storing and accessing computer data binders weighing up to 30 pounds each, and moving small office equipment such as FAX machines, telephones, and typewriters. Therefore, even those employees whose duties do not normally require bending and lifting should be trained in proper lifting technique and educated in basic back physiology.

There are basically two types of back injuries: those that occur to the spinal vertebrae, and those that occur to the back muscles and tendons.

The spinal column consists of a series of bones called *vertebrae*. The spinal cord is a bundle of nerves originating in the brain which pass through the center of the spinal column. Some of these nerves exit the spinal column at each vertebrae, eventually terminating at various locations throughout the body.

The brain, acting as a central processing unit, controls the body's operations by sending and receiving signals along these critical nerve pathways. Severe injury to the spinal column can result in an interruption of this information exchange, leading to paralysis or even death.

Located between each pair of vertebrae are the spinal disks. These disks contain a viscous fluid which acts as a shock absorber to the vertebrae, dampening shocks while maintaining appropriate spacing when the vertebrae are flexed as the result of normal activity.

The spinal column, on average, is designed to support a total of approximately five hundred pounds. Exceeding this weight limitation places severe stress on the disks. Aging causes these disks to degrade over time, reducing their shock absorbing ability. Improper lifting technique can accelerate the process. Once a spinal disk has failed, recurrent severe pain is inevitable.

Minor injuries may resolve themselves, given time and rest. However, there will always be a weakness associated with a previously injured disk, and a future proclivity for recurring symptoms.

Obviously, the prudent approach is to avoid injuries through appropriate back care.

Most people doubt that they will ever exceed their back's weight limitation. However, it should be noted that most people do, in fact, approach at the very least the 500 pound limitation, and many exceed the limitation on a daily basis.

If a 10 pound weight is placed at the end of a 10 foot pole, and an attempt is made to lift the weight from the far end of the pole, it will be found that the perceived weight of the 10 pound weight is now 100 pounds. This is a simple law of physics involving levers and fulcrums. A 10:1 ratio exists between the weight at one end of the pole and the lifting force at the other end.

This same 10:1 ratio exists between the top of a person's head and the bottom of their spinal column. Since the upper torso from just below the shoulders to the top of the head, on average weighs approximately 50 pounds, it will be found that a person who bends over at the waist, and then lifts back up again, will be placing at least a 500 pound load on the spinal column. Any amount of weight added to the above movement through improper lifting technique will exceed the normal maximum stress load that the spinal column is designed to withstand and lead to the possibility of severe spinal column injury.

WHEN LIFTING ANY WEIGHT AT ALL, ALWAYS KEEP YOUR BACK AS STRAIGHT AS POSSIBLE, BEND YOUR KNEES TO ADDRESS THE LOAD AND LIFT THE LOAD WITH YOUR LEGS.

The second most frequent cause of back injuries is muscle strain. This is most frequently the result of not warming up before attempting to lift an object. Muscle strain is also often caused by twisting the back while keeping the feet stationary.

WHEN HANDLING A LOAD, ALWAYS STEP IN THE DIRECTION THE LOAD IS TO BE PLACED. NEVER TWIST THE UPPER TORSO WITH A LOAD WHILE THE FEET ARE STATIONARY.

Muscle fibers are very similar to plastic material; when cold the fibers are very brittle and can tear or break easily. When warm these fibers are very stretchable and flexible.

ALWAYS WARM UP BEFORE PERFORMING ANY TASK WHICH REQUIRES MUSCULAR FLEXIBILITY.

Doing light calisthenics before beginning work each day, and again after a long break, will be very beneficial to avoiding strain injuries.

It is also important to wear the appropriate clothing to ensure that muscles and tendons are kept warm throughout the work period. Layers of light clothing are preferable to one heavy article of clothing. Wearing a heavy coat, for example, will lead to rapid warming. When the coat is removed the body cools down rapidly as well, leading to the increased likelihood of a strain injury.

Strain injuries are normally cumulative. Individual muscle fibers which make up the muscle tear little by little over a period of time due to poor lifting technique and/or inadequate warmup procedures. Finally one day a strain is incurred, and it is presumed that it has occurred as a result of that day's activity, when in fact it occurred over a period of time as a result of repeated micro trauma.

It is the cumulative nature of strain injuries that the workplace wellness program addresses through body movement education and fitness training.

Workers whose duties require frequent material handling are subject to cumulative trauma particularly to the lower back. Frequently the cause of such cumulative trauma is fatigue, rather than improper lifting technique.

Weightlifter belts, when properly and regularly worn by employees, can be very beneficial in reducing cumulative trauma from fatigue. These belts have the added benefit of reminding the employee to keep the torso straight when lifting. If the employee fails to keep the torso straight, the belt will pinch just enough to act as a reminder.

Weightlifter belts, also known as back support belts, come in a variety of colors and materials, and are available from sporting goods stores, wholesalers, and safety equipment distributors. Generally speaking leather belts are the least expensive and the most durable.

Weightlifter belts are generally worn less tightly in industry than when used for competition or for power weightlifting training.

Weightlifter belts have been found to be very useful in reducing work related back injuries. However, it is important for the belt to be worn properly. Belts which are worn too tightly, or worn continuously throughout the day, will lose their effectiveness in reducing low back fatigue.

Always adhere to the following guidelines when wearing weight lifter belts for industrial purposes:

Do not wear the belt too tightly. It should be possible to place your thumb between the belt and your body, and easily slide your thumb around the stomach and hip area.

As work is performed the underlying muscles will become slightly enlarged with additional blood. Normally this additional blood, coupled with the work activity places significant stress on the muscle tendons which attach to the lower back. However, with the belt in place this attendant stress is greatly reduced.

Additionally, the belt will provide a degree of support when a worker lifts a heavy load. However, it must be understood that the belt does not increase the worker's ability to lift weight.

Normally, the belt should not be worn continuously throughout the day. It should be removed at break time, lunch time, and any other time bending and lifting is not anticipated.

The weightlifter belt is not a substitute for the maintenance of a healthy back through a comprehensive education, nutrition, and fitness program.

Generally belts worn in industry are 4 inches wide at their widest point and have single buckle fasteners. Some of the soft weightlifter belts have other types of fasteners, such as touch-type fasteners or pressure buckles rather than a standard belt buckle.

Belts which are 6 inches wide at the back are generally not as comfortable as a narrower belt, but may be appropriate if the worker is very large, has existing low back pain, or simply prefers the extra support.

It may be necessary for the employer to sell the employees on the idea of wearing weightlifter belts. As with any new idea there is often some initial resistance.

Normally there is an initial breakin period of approximately 3 weeks during which time the worker becomes accustomed to wearing the new piece of equipment. After the breakin period most workers wonder how they ever got along without the belt.

Chapter 13

Employee Incentive and Recognition

The time and money invested in the development of a workplace wellness program will be wasted if the workplace population is not encouraged to participate.

It can be anticipated that 20-30% of the employee population will be willing to participate in the wellness programs offered. These individuals, as a rule, are already concerned about their own health, and in all probability have taken steps to maintain their health profile.

The remainder of the workplace population will need to be encouraged to participate through the development of an appropriate incentive and recognition program.

Generally it is this 70-80% of the employee population which possesses the major health risk factors that the workplace wellness program is designed to address.

These are the individuals who smoke, have elevated blood pressure, high serum cholesterol levels, obesity, diabetic proclivities, and other serious health risk factors.

These are the individuals who will avoid participation in wellness activities because of fear, embarrassment, or lack of understanding in regard to the risks that they are facing through nonparticipation.

Frequently self-esteem perceptions are lower in these individuals, and hence they are unlikely to participate in self-improvement programs on their own.

These individuals are also the primary health care users within the employee population. These are the individuals who file workers' compensation claims, who are most frequently absent from work, and are therefore the less productive employees.

In developing a wellness program, thought must be given to the anticipated objections which will be raised by the non-health-conscious portion of the workplace population. Interest surveys should be utilized to uncover these objections, and then incentives designed to overcome the objections which are anticipated.

A very important consideration will be the confidentiality of both health assessment data and interest survey data. Assurances of anonymity are important in attempting to overcome the anxieties of individuals who have not previously participated in wellness activities.

Paradoxically, recognition for success is the best way to overcome feelings of low self-esteem. Therefore, once an individual overcomes the initial anxiety about participating in the wellness program, his or her achievements must be rewarded and recognized in order to positively reinforce the gains made toward optimal health.

Incentive awards can be monetary, but monetary reward is not essential and frequently not as effective as peer recognition. For example, the giving of low cost gifts, such as pens, T-shirts, caps, jackets, pins, or patches can be very effective as recognition of success.

However, the relationship between the gift item and the achievement is very important. For example giving a jacket for achievement of the first goal, followed by patches indicative of additional goal achievement, directly links the accomplishments to the task. Of great importance is that all the employees recognize the progress and achievements made by themselves and their peers. This approach generates tremendous self-esteem and esprit de corps.

However, it is important that goal levels be set conservatively. No goal should ever be unattainable by any member of the workplace population. Achievement recognition must not become elitist.

Therefore, if a wellness program includes team and individual sports it is important for those recognition awards to be completely different from and no greater than those given for achievement of wellness goals.

Recognition awards should be established for individual activity achievement as well as attainment of overall wellness goals. There is nothing wrong with having higher cost awards for overall achievement, with smaller awards for activity achievement.

This is not to indicate that expensive gifts are mandatory. Peer recognition and support are the key factors.

For example, if a walking program is developed, recognition awards should be established for miles walked, or time involved in the program, as well as specific reduction of risk factors. Remember, however, that confidentiality must be maintained in regard to specific medical data. Overall goal attainment should be based upon ranges of achievement, i.e., cholesterol levels between 200 and 240 might constitute one level, with lower ranges down to an optimal level of cholesterol. Or recognizing an individual percentage of change.

"Touch" patches (exemplified by the brand name Velcro) which can be removed and replaced at successive achievement levels are ideal for this type of recognition approach.

The wellness director should designate key wellness coordinators to maintain achievement logs for program participants. Additionally, logs should be kept by instructors in regard to goals reached within individual programs. Again, where individual health risk factors are tracked, i.e., blood pressure, cholesterol, weight, and so forth, confidentiality must be maintained.

The wellness director should establish incentive guidelines and monitor achievement progress to ensure that employees are recognized and rewarded consistently and in a timely fashion.

The key word in incentive programming is *recognition*. A newsletter is an effective means of companywide recognition. Depending upon workplace population size, a newsletter dedicated to wellness might be appropriate. Individual employee profiles and achievements can be highlighted in each issue. Further, the newsletter can also serve as an educational and information reinforcement tool. Newsletters can be used to publicize the wellness program and attract new participants. This is especially effective when nonparticipating employees see their peers successfully profiled.

The goal of any employee incentive and recognition program is to attain and retain participation by the entire employee population. Incentive programs need to be flexible, with modifications made as needed to encourage employees to remain in the wellness program for the tenure of their employment.

To summarize, the motivational aspects of the wellness program cannot be overemphasized. The wellness director must develop a high-quality incentive and recognition program to attract maximum participation in the wellness program.

Motivational considerations should not be an afterthought, introduced only when the program is underway and participation is found to be less complete than anticipated. Nor should motivation be left until participation begins to wane. It is much more difficult to redevelop interest than it is to stimulate and maintain interest in the initial program.

Appendix 1

Survey Questions

The following surveys are presented for informational purposes only. We strongly advise that anyone intending to use these, or any other surveys for the purposes of establishing a wellness program, consult with legal counsel well versed in labor, privacy, and employment law.

INFORMED CONSENT:
EXERCISE PARTICIPATION

The Undesigned hereby voluntarily consents to engage in a fitness class.

(Employer), which subscribes to the American College of Sports Medicine guidelines, recommends that you see a physician before beginning this and any exercise program.

If you are pregnant, or if you suspect that you may be pregnant, (employer) strongly recommends that you see a physician specializing in childbirth (Obstetrician). Early pregnancy care is very important to your own health, and to the health of your baby. Do not start this or any exercise program without the approval of your physician.

The undersigned releases and discharges (employer) and its officers, agents, managers, staff, technicians, and other connected therewith from all claims or damages whatsoever that the undersigned or his/her representatives may have arising from, or incident to this test.

Furthermore, the undersigned may participate in physical conditioning programs and/or use any of (employer's) facilities, services, or equipment, at his/her own risk. (Employer) shall not be liable for any damages for personal injuries sustained by anyone in, on, or about the premises while participating in such physical conditioning programs. The undersigned does hereby release and discharge (employer) from any and all claims, demands, or actions, arising out of the use or intended use of its facilities, services, or equipment, including, without limitation, any claim for personal injuries, resulting from or arising out of the negligence of (employer), its officers, agents, employees, or any other person.

It is especially important that you see a physician if you have, or suspect that you have any of the following:

Diagnosed high blood pressure (greater than 140 mm Hg/90 mm Hg)
Serum cholesterol greater than 240 mg/dl
Smoke tobacco products
Diabetes mellitus
Family history of coronary (heart) or other atherosclerotic disease in parents, or siblings prior to age 55
Pain or discomfort in the chest, or surrounding areas
Unaccustomed shortness of breath, or shortness of breath with mild exertion
Dizziness
Ankle edema (swelling)
Palpitations or tachycardia (heart fluttering)
Known heart murmur

Again, if you have any of the above symptoms, or develop an unusual response to exercise, or you are pregnant, see your physician.

I have read this form, and understand the contents herein:

Employee's signature: Date:

INFORMED CONSENT: FITNESS PROFILE

The undersigned hereby voluntarily consents to engage in a health/fitness profile evaluation of:

Aerobic capacity—You will perform a submaximal exercise assessment on a cycle ergometer, or treadmill. The exercise intensity will begin at a level you can easily accomplish and will be advanced in two minute stages until you reach 60% to 80% of your maximum heart rate, depending upon your current age and fitness level. We may stop the test, or you may stop the test at any time because of fatigue, discomfort, or other appropriate reasons.

Flexibility—You will perform this test in a sitting position on the floor reaching toward your toes. If you have any back problems, or other joint injuries, we may choose, or you may choose to exclude this test.

Resting heart rate and blood pressure—This measurement is taken before you begin any of the above evaluations.

Body weight—Your weight will be measured on a typical scale.

Body composition—Skinfold measurements will be taken at various sites on your body so that we can calculate your body fat percentage.

Total cholesterol—Several drops of blood will be taken by pricking the end of your finger on your nondominant hand.

There are certain changes that may occur during the evaluation process, including abnormal blood pressure response, fainting, heart rate disorder, and in extremely rare instances heart attack, or death. Every effort will be made to minimize these risks through preliminary examination. (Employer) subscribes to the American College of Sports Medicine guidelines which state that for anyone "at or above age 45, it is desirable for these individuals to have a maximal exercise test before beginning exercise programs."

As a result of this evaluation, (employer) will be better able to decide what type of physical activities you may engage in with little or no hazard. Any questions about the procedure are encouraged at any time. Your permission to perform this evaluation is voluntary and you are free to deny consent for any part or all of the evaluation at any time.

The undersigned releases and discharges (employer) and its officers, agents, staff, faculty, technicians, and any other connected therewith from all claims or damages whatsoever that the undersigned or his/her representatives may have arising from, or incident to this test.

Furthermore, the undersigned may participate in physical conditioning programs and/or use any (employer) facilities, services, equipment at his/her own risk. (Employer) shall not be liable for any damages for personal injuries sustained by anyone in or about the premises. The undersigned does hereby release and discharge (employer) from any and all claims, demands, or actions, arising out of the use or intended use of its facilities, services, or equipment, including, without limitation, any claim for personal injuries, resulting from or arising out of the negligence of (employer), its officers, agents, employees, or any other person.

Client Signature: Date:

APPOINTMENT NOTIFICATION

Welcome to (Employer) Wellness Program

You are scheduled for your health/fitness assessment on: _____ at _____. This assessment involves a series of tests that helps us identify your current health/fitness level by measuring the following:

Resting heart rate and blood pressure
Total cholesterol
Body fat percentage
Strength
Aerobic capacity

There is no need to change your daily routine or diet except to avoid drinking caffeine prior to your assessment. Caffeine will influence your resting heart rate and blood pressure measurements.

Dress casually in comfortable clothes such as T-shirts and tennis shoes. Wear shorts if possible so that a skinfold measurement of your thigh can be made.

The cost of your health/fitness assessment is being paid for in its entirety by (employer) as part of an ongoing concern for the well-being of our employees.

The health/fitness assessment is fun and easy. If you have any questions about medical considerations we can reschedule you. We (employer) are billed for all missed appointments.

HEALTH RISK APPRAISAL AND LIFESTYLE PROFILES

Personal History

(Please write your answers at end of each sentence. If more space is needed use the reverse side of this page.)

1. How long has it been since your last complete physical examination?
2. Have you ever been told that you have diabetes?
   If yes, before age 40?
   If yes, after age 40?
3. Have you ever had a heart attack?
4. Have you ever had (please indicate YES or NO after each item):
   Heart surgery?
   Bypass surgery?
   Angina (chest pains)?
   Angioplasty?
   Stroke?
   Blood vessel surgery?
5. What is your usual blood pressure?
   Over 140/90 mm Hg
   Around 120–140/90 mm Hg
   About 120/80 mm Hg
   Below 120/80 mm Hg
   I do not know my usual blood pressure.
6. Would you participate in a blood pressure screening if offered?
7. What is your serum cholesterol level?
   More than 240 mg/dl?
   Around 200–239 mg/dl?
   Below 200 mg/dl?
   I do not know my serum cholesterol level.
8. Would you participate in a serum cholesterol screening if offered?
9. Have you ever had, or been diagnosed as having any of these symptoms, or conditions:
   Coronary heart disease
   Heart trouble of any kind
   Pains in your chest

Feel faint, or have spells of dizziness
Stroke
Cancer
Ulcerative colitis (stomach ulcers, or pain)
Emphysema
Anemia
A bone, or joint problem that may be aggravated or become worse with exercise
Chronic cough or hoarseness
Unplanned weight loss of more than 10 pounds in the past two months
Shortness of breath
Abnormal pulse (rapid, or irregular)
Cirrhosis of the liver
Is there any physical, or mental condition that you would like to tell us about? If yes, please use reverse side.

10. How recently have you had each, or any of the following screening tests? (Please give approximate date after each applicable item.)
Cholesterol
Blood glucose
Blood pressure
Resting EKG
Chest X-ray
11. Would you participate in a health screening if offered free of charge?
12. Your physicians name?

Family History

13. Have any of your relatives had any of these conditions?
Cancer
Diabetes
High blood pressure
Coronary heart disease
If yes, please indicate type of relative, grandparents, parents, uncle, etc.
14. Within your immediate family, parents, sisters, brothers, has anyone had a heart attack or by pass surgery:

No
Yes, before age 60.
Yes, after age 60.

Lifestyle

15. Which of the following best describes your eating habits (circle or cross out as appropriate):
    One serving of red meat, or fried foods per day
    More than 7 eggs weekly
    Red meat 4 to 7 times per week. Some margarine, low fat dairy products, cheese and/or fried foods.
    Poultry, fish, little or no red meat. Some margarine, skim milk, low fat dairy products.
    An ovo-lacto-vegetarian. No meat, but eggs and milk products are included in diet.
    Strict vegetarian. No meat, dairy or animal byproducts.
16. Would you participate in nutritional education classes if offered?
17. Would you participate in a weight management program if offered?
18. What is your overall level of exercise activity?
    Trainer: 5-7 times per week, training for a specific athletic event.
    Vigorous: 5-7 times per week. 5-10 hours per week at an intensity similar to strenuous jogging. Include heavy labor or weight training.
    Weight training: done only for body sculpting.
    Calisthenics: for body toning.
    Maintenance: 3-5 times per week, 1-5 hours per week.
    Low intensity.
    Low: Less than 3 times per week. Less than 1 hour per week.
    Sedentary: No regular exercise involvement.
19. Would you participate in fitness testing if offered?
20. Which activity programs would you participate in if offered:
    Exercise skills
    Walking
    Weight training
    Stretching
    Team sports (baseball, volleyball, basketball, etc.)

Calisthenics
Strength building
Yoga
Running
Martial arts
Swimming
Cycling
Dancing
Back Care

21. Is there an activity which you would especially like to see offered?
22. Which group(s) do you most readily identify with?:
Feeling rushed.
The word "stressed" would describe how you often feel.
Daily headaches.
Frequent neck and shoulder aches.
No daily time available for self and self-care.
No relaxation time during the week.
23. How many alcoholic beverages do you drink normally each week?
24. Do you smoke tobacco?
Cigar, Cigarettes, Pipe?
How many per week?
25. Have you ever smoked?
26. When did you stop smoking?
27. How old were you when you started smoking?
28. Did one or more of your parents smoke?
29. Does anyone in your household currently smoke?
30. Would you participate in smoking cessation classes if offered?
31. Would you like your spouse or household companion to participate in smoking cessation classes as well?

EMPLOYEE INTEREST SURVEY

Name:
Work Phone:
Date:

Are you interested in participating in a workplace wellness program?

If no, why? (Use reverse side of form.)

In your opinion, which of the items listed below would be important to include in a workplace wellness program:

Physical fitness activities
Nutrition education
Weight management
Health education
Stress management
Relaxation technique
Chemical dependency education
Hypertension screening (blood pressure)
Cholesterol screening
Recreational activities

First aid
Safety awareness courses
Smoking cessation
Exercise skill building
Back care program
Employee assistance (elder care, child care, finance management)
Health appraisal
Fitness testing

Which of the following activities would you participate in if offered:

Walking
On-site massage therapy
Weight training
Rope skipping
Body fat management
Calisthenics
Exercise classes
Healthy cooking

Yoga
Stress management support group
Martial arts
Swimming
Stationary cycling
Dancing

What activity would you especially like to see offered?

Please give us any additional information you feel would help us to design a program to meet your needs. (Use reverse side of form, or additional paper if needed.)

Do you have any special skills or knowledge which you feel would be helpful to the development or success of a workplace wellness program?

Would you be interested in learning to be a program coordinator or wellness team leader?

SCREENING EVALUATION

Type of screening:

Your comments will help us to improve our programs. Please take the time to fill out this form.

1. Was the screening run smoothly?
2. Do you feel the screening was efficient?
3. Were the screening personnel friendly?
4. Did the screeners explain the purpose of the test and your test results clearly?
5. Did the screeners answer your questions to your satisfaction.
6. How do you feel the screening could be improved?
7. Have any questions related to the screening come up in your mind since the screening? (Use the reverse side. Include your work telephone number.)
8. Which of the screenings listed below would you also like to see offered?
    Blood pressure
    Physical fitness assessment
    Exercise stress test
    Diet analysis
    Blood cholesterol
    Colorectal cancer
    Breast cancer
    Uterine cancer
    Other cancer (specify)
    General health risk
    Mouth cancer
    Vision
    Glaucoma
    Diabetes
    Blood analysis

Please return to:

INTEREST SURVEY

If a workplace wellness program were offered, which of the following areas would you most likely to participate in:

Cardiovascular fitness
Personal stress management
Organizational stress management
Smoking cessation (how to quit smoking)
Weight control
Nutritional education
High blood pressure
Medical self-care education
Alcohol/drug use education
Mental/emotional problem resolution (depression, anxiety)
Parenting skills
Marital problems
Assertiveness training
Educational/career planning
Spiritual values
Philosophical values
Interpersonal communication skills (getting along with people)
Home budgeting and finance planning
Automobile safety
Time management

When would you most likely attend?

After work
Lunch
Evenings
Before work

Would you most likely attend (list activity next to choice):

Once
6–8 weeks
Continuously

Would you be interested in programs that would include members of your family?

Are there changes you would like to see made in your work environment that would better support healthy behaviors? (Use reverse side of form as needed.)

Would you be willing to assist in developing and leading a program?

What programs would you be willing to assist in?

What programs would you especially like to see developed?

Name:
Department:
Work Telephone:

WELLNESS SURVEY

1. Male _____ Female _____
2. Please circle your age group: 15–21   21–30   31–40
   41–50   51–60   60 +

Please indicate YES or NO after the following statements (3–8):

3. I sometimes exercise.
   I exercise at least once per week.
   I exercise vigorously for more than twenty minutes at least three times per week.
4. I often eat breakfast.
   I usually have two meals per day.
   I normally eat three meals per day.
   I often eat in a rush, and frequently skip meals.
   I avoid excess fat in my diet.
   I consciously eat high fiber foods.
   I use salt on my foods to improve its taste.
5. I would like to lose weight.
   I am happy with my weight.
   I would like to gain weight.
6. I frequently have trouble sleeping at night.
   I need to take a nap each day.
   I get insufficient sleep two, or more nights per week.
   I sleep well nearly every night of the week.
7. I smoke tobacco products—cigarettes, cigars, pipe.
   I drink alcohol on a daily basis.
   I avoid drinking too much caffeine.
   I regularly use prescription tranquilizers, or similar prescription drugs.
8. I feel comfortable with self care for minor medical problems.
   I have had back pain on and off during the past six months.
   I practice stress management and coping techniques when necessary.
   I see a doctor anytime I think I might be getting sick.

9. Please indicate any health concerns you have about yourself, or your family. (Use the reverse side of this form.)

10. Would you like to participate in a wellness program if it were offered?
11. Please put a check mark next to the wellness activity in which you would like to participate:
    Aerobic exercise
    Weight management
    Smoking cessation
    Health screening
    Coping with stress
    Alcohol/drug use education
    Accident prevention
    Safety education
    Parenting
    Walking
    Blood pressure screening
    Cholesterol screening
    Cancer screening
    CPR training
    First aid training
    Wellness education
    Nutrition education
    Weight training
    Financial planning
    Weight management
    Back care education
    Medical self-care education
    Retirement planning
12. Please indicate what subject you would especially like to see included in a wellness program.
13. Please indicate the time of the day, and the days of the week you would be able to participate in a wellness program.
14. Are there any special skills that you have which you would like to share with other employees in a wellness program setting?
15. Would you be willing to be a wellness group leader?

INTEREST SURVEY

Please indicate numerically after each item, on a scale of 5, your interest in participating in the following programs.

Appendix 1: Survey Questions   123

1 = not interested - 5 = very interested

Self-help health and medical education
Quit smoking
Nutrition education
Stress management
Aerobic exercise
Weight training
On-site massage
Dancing for fitness
Accident prevention
First aid training
CPR training
Team sports (circle): Basketball, softball, volleyball, other

Would you like your family to participate in the wellness program also?

What wellness related program would you especially like to see offered?

ENVIRONMENTAL AUDIT

On a scale of 1-5, please indicate your satisfaction with the following work environment items. All answers are confidential.

1 = not satisfied - 5 = very satisfied

Lighting in your work area
Comfort of your workstation
Noise level
Temperature
Adequate fresh air/ventilation
Choice of food in cafeteria
Food offered in vending machines
Supervisor handles conflict situations adequately
Work schedule
Priority of work completion
Input in priority and schedule decisions
Overtime
Housekeeping
Safety awareness
Accident prevention
Equipment maintenance
Adequacy of tools and equipment with which to work

Please indicate any areas of change which you would especially like to see accomplished.

MANAGEMENT SURVEY

Please answer YES or NO after each item. (Use the reverse side of this form if needed.)

Workplace Wellness

1. Employees need accurate information on the following:
   Health risks
   Accident prevention
   Safe operating procedures
   Health care costs
   Accident costs
   How to change bad habits
2. People will change their behavior informed, given incentive for motivation, and supported.
3. Healthy people are more productive than unhealthy people.
4. My work environment has an impact on my health.
5. If a wellness program were instituted I would expect to see the following changes in myself, or my co-workers:
   Participation in a regular exercise program.
   Reduction in tobacco smoking.
   Greater ability to cope with daily stress.
   Better weight management.
   Better eating habits, lower cholesterol, lower blood pressure.
   Reduction in use of caffeine, sugar, salt.
   Reduction in misuse of drugs and alcohol.
6. Are you looking forward to the implementation of a workplace wellness program?
7. Would you personally participate in a workplace wellness program?
8. What special skills do you have which you would like to share with other employees in a wellness program?
9. Would you be willing to be a wellness team leader?
10. What program would you especially like to see included in a workplace wellness program?
11. Would you like to see family member participation as well?

Appendix 2

Methods of Tracking Costs and Costs Savings

Example

1. Health Insurance Premiums for the past five years.

Year 1. 1986 = $ 125,000
Year 2. 1987 = $ 132,500
Year 3. 1988 = $ 143,100
Year 4. 1989 = $ 157,410
Year 5. 1990 = $ 176,299

Average percentage change between years 1 and 5 = 17%.
Assuming a constant rate of change, by 1995 we can expect to pay $206,269 for the same number of employees, an increase of $29,970 in the next five years.

2. Disability Payments.

Workers' compensation: Current year total:	$100,000
(100 employees) Average per employee:	$1,000
Health plan disability: Current year total:	$75,000
Average per employee:	$750
Total per employee:	$1,750

3. Absenteeism.

Total days of absenteeism:	175 (1.75 days per employee)
Absenteeism times total per employee:	$306,250
Insurance Coverage for Health related costs:	
Health Insurance Premium:	176,299
Workers' Compensation Insurance Premium:	75,000
Total:	251,299

Total absenteeism costs: $306,250
Total insurance premiums paid: $251,299

Shortfall = $54,951.00 will require increase in premiums in coming year to correct.

Or, insurance premium per employee = $251,299/100 = $2,512.99.

Paid claims per employee = $3,062.50

In most instances health care cost increases are paid directly from the employer's profit because the pricing of products or services produced by the employer cannot be increased enough to offset the increases in health care costs while still remaining competitive in the market place.

The only solution to this dilemma is reducing health care utilization through the implementation an effective workplace wellness program and an effective loss control program.

If the shortfall of $55,000 noted above were invested in effective workplace wellness and loss control programs, the result would be an investment in future earnings, rather than simply paying out more and more dollars to insurance premiums.

By tracking actual health care costs an employer can anticipate future premium increases, and also gain some insight into what would be a reasonable investment in the creation of an effective workplace wellness program.

As the cost reducing effects of workplace wellness and loss control manifest themselves, the investment made in budgeting for these programs is returned to the employer through reduced insurance premiums as the result of reduced health care utilization. This offset does not take into consideration the increased profits realized through increased production.

While morale and production improvements will be seen immediately as the result of implementing effective wellness and loss control programs, reduction in insurance costs will take at least two years, and probably three years to be realized.

The decision to implement a wellness and loss control program should be geared to a long term consistent plan.

FREQUENCY RATES

Frequency rates are based upon 100 employees working 40 hours per week for 1 year. In developing the formula statisticians have made

Appendix 2: Methods of Tracking Costs and Costs Savings

adjustments to allow the applicability of the formula regardless of the number of employees employed by any given employer.

The basic formula is: the number of claims incurred multiplied by 200,000 and then divided by the actual manhours worked by all employees for any given employer.

The frequency rate formula may be used to track health benefits utilization.

Example

15 employees
3 claims
35,000 manhours worked (includes 5000 hours of overtime)

3 × 200,000 = 600,000 divided by 35,000 = 17.1

Frequency rates alone do not indicate how effective a wellness program is in reducing health benefits utilization, but they are useful in demonstrating relative trends.

Frequency rates are useful as a wellness program tool when they are used to track claims over a five year or longer period of time.

They are also useful as an adjunct to other data when developing the initial Wellness budget. The frequency rate for the prior five years can show the health benefits utilization trend prior to Wellness program implementation.

And they are useful after program implementation as another means of tracking progress.

Example

1986. 11 employees. 3 × 200,000 / 24,200 = 24.7
  3 claims
  24,200 hrs (includes overtime)
1987. 13 employees. 5 × 200,000 / 28,600 = 34.96
  5 claims
  28,600 hrs
1988. 15 employees. 4 × 200,000 / 33,000 = 24.24
  4 claims
  33,000 hrs (includes overtime)

1989. 17 employees. 7 × 200,000 / 34,000 = 41.17
   7 claims
   34,000 hrs
1990. 15 employees. 6 × 200,000 / 30,000 = 40.00
   6 claims
   30,000 hrs

As can be seen from the above example this employer has a trend of generally increasing claims frequency relative to the number of manhours worked.

The number of manhours worked is the determining factor, not the number of employees.

It should be noted that some years contain no overtime, whereas others contain overtime. It is important when tracking frequency rates to separate overtime from the formula if any year used has zero, or nearly zero overtime.

Let us adjust this example to zero overtime for all years:

1986. 11 employees. 3 × 200,000 / 22,000 = 27.27
   3 claims
   22,000 hrs
1987. 13 employees. 5 × 200,000 / 26,000 = 38.46
   5 claims
   26,000 hrs
1988. 15 employees. 4 × 200,000 / 30,000 = 33.33
   4 claims
   30,000 hrs
1989. 17 employees. 7 × 200,000 / 34,000 = 41.17
   7 claims
   34,000 hrs
1990. 15 employees. 6 × 200,000 / 30,000 = 40.00
   6 claims
   30,000 hrs

While the differences between overtime and no overtime in this particular example are minimal, it can be seen that when overtime varies significantly from year to year the overtime discrepancies must be factored out of the monitoring calculation to ensure consistency in tracking program needs, or results.

When compared we find:

	1986	1987	1988	1989	1990
With overtime:	24.7	34.9	24.2	41.7	40.00
Without overtime:	27.2	38.4	33.3	41.7	40.00
Employees	11	13	15	17	15

Frequency rates can also be displayed as a graph to dramatically show trends.

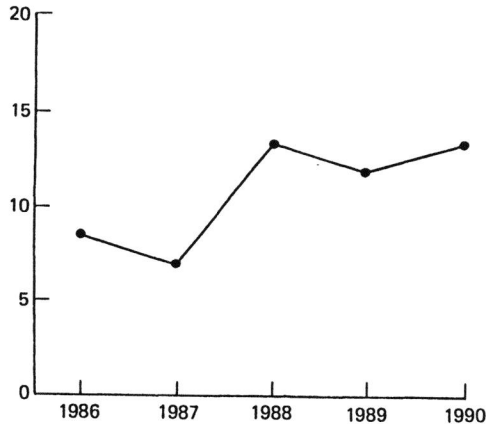

VARIANCE CHARTS

Another useful tool for tracking wellness trends by department is the variance chart. This chart or graph can be used for daily, weekly or monthly display of claims. A chart of this type can also be used to correlate conditions which may be contributing factors to claims.

Dept	Jan	Feb	March	April	May	June	July	August	Oct	Sept	Nov	Dec
A	x	x	x				x			x		
B	x	x						x				
C	x		x									
D												
E					x							
F					x					x		x

For instance, in the above example the largest grouping of points occurred during the coldest months of the year. If the corresponding departments involved were exposed to cold weather perhaps the cold environment was a contributing factor.

Similar charts can track types of claims, body parts involved, types of ailments, and other relevant data.

WELLNESS PROGRAM CORRELATION CHART

The chart of greatest value to the employer will correlate wellness efforts with statistical data. Such a chart might include claims data and wellness module attendance data.

TABLE A2-1.   Wellness Program Correlation Chart.

Month	January			
Week	1	2	3	4
Dept A	A	C	A	C
B	A			
C	A	C		
D			A	A
E				A
F		A	C	

C = Claim
A = Attendence

A chart of this type if continued through out the year with annotations describing the types of claims submitted and the wellness modules participated in can be very valuable in planning the next year's wellness efforts.

The chart can be adapted for tracking a number of variables and correlating their importance to the results of the employer's wellness efforts.

THE ONE CARD SYSTEM

The one card system can be used to suspense date all routine wellness activities.

The one card system consists of a file card box, file cards, and file

Appendix 2: Methods of Tracking Costs and Costs Savings

card dividers showing events and/or months of the year, 1st, 2nd, 3rd and 4th weeks of the month and/or can be further broken down into the days of the week and/or times of the day.

The file cards may be used to list wellness activity modules and/or an employees' scheduled participation in such modules.

Using the one card system, a card would be made up for each event indicating the type of event, individual involved, and/or goals to be accomplished. Other pertinent notes can be added to the card.

The card is then inserted in the appropriate divider slot. It can now be forgotten because when the date/time comes up it will be retrieved as a reminder of the required event.

At the beginning of each week the person responsible for coordinating the events will pull all of the cards for the week and schedule the events and/or employees into the appropriate time/day slots.

Following the event, additional notes of reminder for the next occurrence of the event can be added to the card. Using a pencil facilitates the addition and subtraction of notes following an event. The card is then moved forward into the appropriate slot for the next occurrence of the event and so the system continues to operate for all events indefinitely.

Those employers with appropriate software can establish a similar system on their computer. Suspense date software is also available for personal computers. However, for the small employer such software is often more time consuming to update than the one card system described.

The primary advantage of the one card system or a similar software program is that it removes the need to remember the dates of events and it provides continuity to the entire program regardless of any personal changes which may take place.

Often an employer establishes an excellent wellness program which is competently run by one individual. However, subsequent to that individual's leaving the employer the program completely falls apart because the person replacing the former wellness coordinator has no idea of what was being accomplished, or what was scheduled to be accomplished in the near or distant future.

The one card system or similar software eliminates this continuity problem because the next responsible person need only thumb through the file to get a good working picture of those routine wellness events that need to be accomplished.

Appendix 3

Fitness Training

Weight training is an important aspect of physical conditioning. It helps an individual to maintain the strength necessary for lifting everyday objects, and for the maintenance of muscle tone for work and recreational needs.

According to the American College of Sports Medicine, strength maintenance is an important adjunct to an aerobic conditioning program.

There is a variety of weight training equipment which can be used for this portion of the wellness program. Generally, weight training equipment is divided into four categories:

1. Free weights
2. Infinite resistance machines (IR)
3. Variable stacked weight machines (VSW)
4. Variable resistance stacked weight machines (VRSW)

Free weights have the advantage of low cost and low maintenance. They are very portable and easily incorporated into an on-site program. Free weight exercise trains the finer muscles as well as the large muscle groups and allows a full range of motion. One of the disadvantages of virtually all machine weight training systems is that they are designed for an average range of size. A very large or small person may have difficulty training through the full range of motion when using weight machines.

Free weights have the disadvantage of requiring extra care to avoid injury. For this reason it is important for participants to be well supervised when utilizing free weight training. The buddy system should be used when training with free weights to ensure that someone is standing by to assist if needed.

Free weights should be purchased in the smallest available increments of weight to ensure that an adequate range of weights is available for all participants.

Infinite resistance machines create the perception of lifted weight through the use of air pressure, hydraulic resistance, or the flexor qualities of the material used in the machine's resistance design, such as rubber or spring metal.

One of the advantages of IR machines is the nearly unlimited amount of progressive weight which can be simulated. An equivalent amount of stacked weight would require many more individual weights and much longer cycles of movement.

With the exception of air or hydraulic pressure machines, the major disadvantage of IR machines is that the initial force required to begin the exercise is always very light. This condition tends to deemphasize the development of the weakest portion of the muscle, while emphasizing the development of the strongest portion of the muscle. This intrinsic characteristic can lead to future injury if these machines are used exclusively for weight training.

Variable stacked weight machines have the advantage of isolating muscle groups when an exercise is performed, as well as allowing the user to operate the machine alone without the exposure to injury from a falling weight.

The greatest disadvantage of VSW and all other machines is that the trajectory of motion and extent of travel is entirely controlled by the design of the machine.

Variable resistance stacked weight machines are best exemplified by the patented Nautilus system. The machine is designed with a cam counterweight which offsets the total weight being moved. As the exercise progresses the weight becomes progressively heavier until it reaches maximum resistance during the strongest portion of the muscle's exercise cycle. Then the counterweight reduces the loading as the muscle continues toward full contraction. The VRSW machine therefore allows an individual to work with the maximum amount of weight possible for the strongest portion of the muscle while still applying appropriate resistance during the weaker stages of an exercise.

No one system is perfect for all training situations. Therefore most training facilities use a combination of these equipment types. Should an employer elect to establish a weight training facility on site, it would be advisable to retain the services of a health club design consultant if the employer's staff does not have the requisite expertise to design a complete weight training facility.

When developing a weight training program it is important to include the female employee population in the early planning stages. Subsequently, in order to reduce the anxiety females often have in regard to lifting weights, the program should include a females only day or class.

Educating the employee population on the importance of strength training will help to encourage involvement in the weight training program. Education classes will inform prospective participants of the benefits of strength training, which include a greater ability to perform daily tasks, reduced risk of heart disease, and the improvement of body composition through muscle gain.

SAFETY TIPS FOR WEIGHT TRAINING

Require that a trained spotter always be available in the weight room when free weight exercises are being performed. Encourage the buddy system of training.

Utilize machines for heavy lifting exercises. Make certain that all machines have built in safety features to eliminate the possibility of a weight being dropped onto an exercise participant or training buddy. Of particular concern are the following exercises: squats, calf lift, bench press, military press.

Ensure that all participants are trained and supervised in proper lifting techniques for each exercise. Supervise and control the rate of increase and the amount of weight used. Develop a variable phase training program which allows tendons and ligaments to catch up to muscle development.

Insist that all participants wear closed toe shoes. Several manufacturers market shoes designed for weight training; while these shoes are well designed they are not essential to the program.

Ensure that all participants use weight clamps to retain the weights on the weight bars. Even when Olympic size weight bars are in use, retention devices are mandatory. A falling weight is a danger to the participant lifting as well as to spotters and other participants. Flooring and equipment may also sustain expensive damage from a falling weight.

Encourage the use of weightlifter's belts, especially for heavy lifts and for overhead lifts.

Weight should always be added gradually and in no more than 5 pound increments within the varied phase training routine.

Train participants not to work the same muscle groups two days in a row. Muscle recuperation varies from person to person due to age, stress, and body composition factors. Participants should be trained to "listen" to their bodies. Some training sessions seem easier than others; participants should not try to out do themselves at every session. Participants who push themselves through fatigue to perform at prior peak levels will eventually sustain serious injury.

In choosing equipment for on-site programs it is best always to choose high quality devices designed for commercial gymnasium operations. While such equipment will initially be more expensive than equipment designed for home use, it will in fact outlast, outperform, and require far less maintenance than lighter gauge models.

Regardless of the quality of equipment purchased some maintenance will be required. If the employer's operation does not include an equipment maintenance department, then a maintenance contract should be purchased. Downtime due to equipment failure will greatly inhibit the effectiveness of the physical conditioning portion of a workplace wellness program.

BASIC NUTRITION

Amino acids are the basic building blocks of every cell in the human body, and are the components of all proteins. Amino acids are derived from the intake of protein found in food as polypeptide molecular chains. The body's ability to break down these chains of protein and extract the amino acids is completely dependent on the efficiency of its digestive fluids and its ability to absorb nutrients through the intestinal tract.

Additionally the body's ability to utilize the extracted amino acids for cellular maintenance and repair is limited by the ratio of certain essential amino acids in any given protein. The quality of a protein refers to the ratio of amino acids contained in and extractable from the protein in question.

There are over 100 amino acids found within the human body; however, only 24 are derived directly from food. The rest are created at the cellular level by the combining of those amino acids derived from food.

The 24 amino acids found in food are: L-alanine, L-arginine, L-asparagine, L-aspartic acid, L-citrulline, L-cystine, L-cysteine, L-glutamic acid, L-glutamine, Glyceine, L-histadine, L-serine, Taurine, L-tyrosine, L-ornithine, L-proline, L-isoleucine, L-leucine, L-lysine, L-methionine, L-phenylalanine, L-tryptophan, L-threonine, L-valine.

The last eight of the above listed amino acids are considered essential. Some disagreement exists among researchers as to whether all 24 amino acids may be considered as found exclusively in food. Therefore the reader will find from 18 to 24 amino acids referred to in nutritional textbooks, depending on the author's educational background.

A great deal has been written concerning the importance of the eight essential amino acids. This tends to give the impression that the other amino acids are not as important and that the body can create them as needed. However, this is not the case at all.

The body synthesizes nonessential amino acids by recycling dead cells from the intestinal tract and catabolizing protein within the body. If the diet is chronically deficient in certain amino acids, or if the diet is deficient in the requisite quantity of high quality protein, this catabolizing reaction will eventually take a deleterious toll on the system. It is for this reason that extreme diets which drastically limit nutritional intake should be avoided.

Recently a great deal of anecdotal information has been disseminated, primarily through a few popular bodybuilding and nutritional magazines, about the efficacy of taking individual amino acids to correct theoretical deficiencies in individual diets or to effect significant change in muscle growth, weight loss, or energy levels. While this approach may be valid where known deficiencies exist, the taking of individual amino acids at random, or through anecdotal prescription, is not a good idea.

In virtually all cases the body requires a combination of three amino acid molecules in their proper relationship to produce the effects observed. An amino acid that has an assigned nutritional role does not necessarily produce the observed effect on its own.

The most prudent approach is to ensure that nutritional intake is appropriate for the workload being placed upon the body both by normal daily activities and by recreational and fitness activities. This is best accomplished through the intake of a well balanced diet.

Some individuals elect to supplement their diets with amino acids to ensure that an optimal level is maintained. Should an individual desire such supplementation, then it is best to take a balanced amino acid supplement rather than individual amino acids.

There are many kinds of amino acid supplements on the market. Some are derived from animal protein, yeast protein, egg protein, or milk protein. All of these types of supplements require digestive enzymes to break down the polypeptide chains and extract the amino acids; therefore, they should be consumed with food.

There is another type of amino acid supplement known as *free form*. These are singular amino acids not linked in polypeptide chains and sold individually, or as formulas based upon ratios found in food, generally that found in egg albumin (egg white).

Free form amino acids are created by breaking down the polypeptide chains through the use of digestive enzymes (also known as pre-digested supplements), or by creating individual amino acids through a fermentation process.

Recently the Food and Drug Administration has required the withdrawal from the market of all fermented or synthetically produced L-tryptophan, including that which would be placed in a balanced free form formula.

Therefore, all free form formulas on the market today contain pre-digested L-tryptophan derived from food sources.

When selecting a free form amino acid supplement the most effective formulas are based upon the ration of amino acids found in egg albumin. This ratio of amino acids has been shown to be the most readily utilized combination of amino acids available to the human body from a food source.

Much of the research conducted on amino acid supplementation was conducted by the National Aeronautic and Space Administration during the *Apollo* program. On some space flights the astronauts consumed only free form amino acids in an egg albumin ratio, with certain vitamins and minerals added. These low residue diets were successful to a limited extent, and proved that it was possible to survive for a limited time with an intake of amino acids and vitamins only. However, long term testing indicated that supplementation was not a viable replacement for a well balanced diet.

Again extreme diets of any type should be avoided. The prudent course to follow is one of an adequate and well balanced diet, with appropriate balanced supplementation as needed.

CARBOHYDRATES, FIBER AND FATS

Carbohydrates may be divided into two basic categories: simple and complex. Simple carbohydrates are sugars including maltose, lactose, galactose, fructose, glucose, and sucrose. The average American consumes approximately 128 pounds of simple carbohydrates per year.

Simple carbohydrates can be used by the body's cells for energy. However, utilization requires specific blood enzymes in combination with insulin. Without these two components the cells are unable to use most sugars. Excessive ingestion of sugar stresses the pancreas gland, which produces insulin, and can in turn lead to health problems associated with sugar intolerance.

Complex carbohydrates include grains, unsweetened cereals, legumes, nuts, all types of vegetables, and foods made from grain products such as bread, pasta, and other baked or grain based foods.

Carbohydrates form a major portion of the diet primarily because of their accessibility from a variety of plant sources, their relative low cost, and ease of storage.

Complex carbohydrates are a source of energy and they also aid in the metabolism of fat. The liver is able to break down carbohydrates and store them in relatively large quantities as glycogen both in the liver itself and in the muscles. When energy is needed for sudden or sustained bursts of activity these glycogen stores are available but rapidly depleted.

In combination with protein, carbohydrates stimulate and repair the immune system, lubricate the joints, aid in growth of bones and muscle, skin, nails, cartilage, and connective tissue.

If carbohydrates intake exceeds the body's ability to store it as glycogen or utilize it for its immediate energy needs, it is then converted to fat and stored by the body as a future source of energy.

The efficiency range of individual utilization of carbohydrates ranges from 50 to 98%. The less efficiently an individual utilizes carbohydrates the more easily the individual converts the carbohydrates to fat. Generally the greater the individual's lean muscle mass the more efficient will be the utilization of carbohydrates and the greater will be the reserve available for glycogen. In any case, as an individual ages carbohydrate utilization efficiency decreases.

FIBER

All vegetables, most fruits, and most grains contain a large percentage of dietary fiber. Fibers such as cellulose and pectin are also carbohydrates. Human beings cannot produce the enzymes needed for breaking down dietary fiber. Therefore, fiber is generally not digested and contributes no caloric energy.

Fiber provides bulk to the intestinal tract, acting to speed the removal of digestive byproducts and residue. Cellulose fiber has the ability to absorb large quantities of water relative to its own weight. Since the contents of the lower intestinal tract are primarily water this fiber is very beneficial in keeping the tract clean.

The transit time required from the point of completed digestion to elimination is decreased by sufficient dietary fiber and correspondingly increased by insufficient fiber. Digestive tract residue contains high levels of digestive byproducts, bacteria, and molecules which can form carcinogenic chemical reactions. For this reason it is very beneficial to ensure that an individual's diet contains a satisfactory quotient of dietary fiber.

FATS

Fats and fatty acids are derived from both protein and carbohydrates. Fatty acids may be contained within certain fats but can also be synthesized by the body, or found in the oils of vegetables and nuts.

Technically fats are part of the lipid group which are composed of carbon, hydrogen, and oxygen just as are carbohydrates. They are a source of energy, as well as being used in certain chemical reactions to repair various tissues in the body.

Dietary fats should be consumed with moderation. The National Academy of science has recommended that no more than 25 grams of dietary fat should be ingested per day.

Dietary fat is found in meats, dairy produce, poultry, and nuts. These sources of dietary fat contribute saturated fatty acids as opposed to unsaturated fatty acids. Saturated fatty acids are metabolized inadequately and are generally associated with heart disease and other serious disorders.

One gram of fat produces nine calories of energy, which is the principal function of fats for the body.

Lipids found in fats and vegetable oils are converted to body fat and stored as a thermal blanket subcutaneously to prevent heat loss. They are also a component of cell membranes which form protective cushions for many tissues and organs.

Generally the diet should not contain more than 30% fat, preferably from unsaturated fatty acids, with less than 10% from saturated fatty acids.

VITAMINS

Vitamins are coenzymes that are used by the body in various chemical reactions. Vitamins are not used directly by the body for energy, but some vitamins are used in the process of energy production.

There are two groups of vitamins, fat soluble and water soluble. Vitamins A, D, E, and K are fat soluble. B Complex and Vitamin C are water soluble.

Fat soluble vitamins are stored in the liver and fatty tissues. Water soluble vitamins are not considered storable by the body. However, when appropriate cellular levels are attained it is possible for the body to maintain a short term reserve for several hours.

If an individual chooses to take vitamin supplements it is important to purchase reputable brands from dependable sources. Except for vitamin A, which should come from natural sources, there is little difference between "natural" and synthetic vitamins. However, "natural" vitamins contain trace amounts of various nutrients included as part of the base source, or as fillers to develop standard tablet size. Nutritious excipients are preferable to fillers with no nutrient value.

Trace minerals are important to enzymatic and electrolytic processes within the body. Most multiple vitamin supplements also contain a number of trace minerals. However, minerals, and most vitamins, require digestive enzymes in order to pass through the lining of the stomach for utilization at the cellular level. Therefore, vitamin and mineral supplements are best taken immediately following or during a meal.

In adding food supplements to the diet it is best to maintain a conservative, balanced formula approach. The stress of modern living, air pollution, and the distance that fresh food travels to market all conspire to cause the body to use up its store of vitamins and minerals while also reducing the levels available from dietary sources. For most individuals this is not a significant problem while for others, in particular those individuals who tend to respond more negatively to perceived stress and/or who maintain a nutritionally deficient diet, food supplementation can be beneficial.

Resources

The following resource guide is for informational purposes only, and does not imply an endorsement by the authors.

HEALTH RISK APPRAISALS

Body Incorporated
P.O. Box 859
San Luis Obispo, CA 93406
(800)488-2415

American Health Management
 and Consulting
85 Old Eagle School Road
Stafford, PA 19087

American Heart Association
7320 Greenville Avenue
Dallas, TX 75231

The Center for Consumer Health
 Education
380 West Maple Avenue
Vienna, VA 22180

Computerized Health Appraisals
13705 Southeast 42nd Street
Clackamas, OR 97015

General Health
1046 Potomac Street N.W.
Washington, DC 20007

The Institute for Personal Health
2100 M Street N.W. Suite 316
Washington, DC 20063

National Health Clearing House
P.O. Box 1133
Washington, DC 20013

Wellness Associates
42 Miller Avenue
Mill Valley, CA 94941

Health Hazard Appraisal
Prospective Health Center
6220 Lawrence Drive
Indianapolis, IN 46226
(317)549-3600

Healthier People
The Carter Center of Emory University
Health Risk Appraisal Program
1 Copenhill
Atlanta, GA 30303
(404)872-2100

Health Path
Stay Well Health Management Systems, Inc.

1285 Corporate Center Drive Suite 100
Eagan, MN 55121
(612)454-3577

Health Plan and Health Plan Plus
General Health, Inc.
3299 K Street NW, Suite 300
Washington, DC 20007
(202)965-4881; (800)424-2776

Health Profile
Johnson & Johnson Health Management, Inc.
One Johnson & Johnson Plaza
410 George Street
New Brunswick, NJ 08901
(800)443-3682

Healthtrac
Blue Shield of California
Two North Point
San Francisco, CA 94133-1599
(415)445-5217

Hope Health Appraisal
International Health Awareness Center
350 E. Michigan Avenue, Suite 301
Kalamazoo, MI 49007-3851
(616)343-0770

Life Scan and Lifestyle Assessment
   Questionaire
National Wellness Institute, South Hall
1319 Fremont Street
Stevens Point, WI 54481
(715)346-2172

Lifestyle Analysis Questionaire
University of Michigan
Fitness Research Center
401 Wastenaw Avenue
Ann Arbor, MI 48109-2214
(313)763-2462

The Lifestyle Assessment System, 2100 HR
Lifestyle Management Reports, Inc.
368 Congress Street
Boston, MA 02210
(617)451-0440; (800)531-2348

Lifestyle Inventory and Fitness Evaluation
Personal Wellness Profile
Wellsource
15431 S.E. 82nd Drive, Suite D
P.O. Box 569
Clackamas, OR 97015
(503)656-7446; (800)533-9355

NUTRITION INFORMATION

Computerized Health Appraisals
Upper Columbia Conference
P.O. Box 19039
Spokane, WA 99219

Medical Data Motion
Southwest and Harrison
Bellview, OH 44811

Pacific Research Systems
P.O. Box 64218
Los Angeles, CA 90064

Bastyr College
144 N.E. 54th
Seattle, WA 98105
(206)523-9585

East Coast Health Foods Organization
1075 Easton Avenue
Somerset, NJ 11747
(201)247-8446

Golden West Nutritional Foods Association
P.O. Box 31040-306

Laguna Hills, CA 92654
(714)831-3549

National Institute of Nutritional Education
1010 Joliet Street, Suite 106
Aurora, CO 80012
(303)340-2054

National Nutritional Foods Association
150 Paulrino Avenue, Suite 285
Costa Mesa, CA 92626
(714)641-7005

Ohio Economic Foundation Association
3185 Trip Road
Bellefontaine, OH 43311
(513)593-1232

Butler Nutrition Associates
89 North Main Street
P.O. Box 924
Kent, CT 06757
(203)927-4005

Colgan Institute
531 Encinitas Blvd #101
Encinitas, CA 92024
(619)632-7722

SMOKING CESSATION INFORMATION

American Cancer Society
90 Park Avenue
New York, NY 10017

American Heart Association
7320 Greenville
Dallas, TX 75231

American Lung Association
1740 Broadway
New York, NY 10019

Medical Data Motion
Soughwest and Harrison
Bellview, OH 44811

Smokers Anonymous
2118 Greenwich Street
San Francisco, CA 94123

STRESS MANAGEMENT INFORMATION

Computerized Health Appraisals
Health Services Department
Upper Columbia Conference
P.O. Box 19039
Spokane, WA 99219

Self-Health Systems
Route One
P.O. Box 127
Fair Grove, MO 65648

PERINATAL RESOURCES

Audio Cassettes

1989 IDEA Prenatal Specialty Course
Info Medix
12800 Garden Grove Blvd., Suite F
Garden Grove, CA 92643
(800)367-9286

Books

Complete Pregnancy and Baby Book
by Vicky Lansky,
Consumer Guide Publications, Intl.
3846 West Oakton Street
Skokie, IL 60076

Marie Osmond's Exercises for Mothers-To-Be and for Mothers and Babies
by Marie Osmond
Maternal and Child Health Center

2464 Mass Avenue
Cambridge, MA 02140
(617)864-9343

Shaping Up for a Healthy Pregnancy
by Barbara Holstein, MS
Life Enhancement Publications
P.O. Box 5076
Champaigne, IL 61820
(800)342-5457

*Swim Lite Prenatal Water Workouts:
Prenatal Water Workout Instructor's
Manual*
by Gretchen Schreiber
Water Warehouse, Inc.
P.O. Box 82404
Phoenix, AZ 85071
(602)863-0018

Organizations/Programs

Perinatal Health and Fitness Network
Dancing Through Pregnancy
Box 3083
Stoney Creek, Branford, CT 06405
(203)481-2200

Moms on the Move Perinatal Training
    Institute
272 Shoreline Highway #3
Mill Valley, CA 94941

Pamphlets

"Pregnancy, Fitness, Post-Birth, Parenting and Infant Care" (catalog)
Penny Press, Inc.
1100 23rd Avenue
East Seattle, WA 98112

"Pregnancy and Exercise"
Melpomene Institute for Women's
    Health Research
2125 East Hennepin Avenue

Minneapolis, MN 55413
(612)378-0545

Products

Leading Lady (body wear, support belt)
24050 Commerce Park
Beachwood, OH 44122
(800)321-4804

Research

Exercise and Pregnancy Study
Melpomene Institute
2125 East Hennepin Avenue
Minneapolis, MN 55413

Videotapes

ACOG Pregnancy and Exercise Series
    (6 tapes)
Feeling Fine Productions
3575 W. Cahuenga Blvd. Suite 440
Los Angeles, CA 90068
(800)621-1203

Babyjoy and
*Marie Osmond's Exercises for Mothers-
    to-Be*
Maternal and Child Health Center
2464 Mass Avenue
Cambridge, MA 02140

Pregnancy Exercise Workout
by Shirley Boyle
4847 Hart Drive
San Diego, CA 92116
(619)563-9331

Pregnant and Fit
by Judith Pettibone
People in Motion, Inc.
1395 Fawnwood Drive
Monument, CO 80132

SENIOR FITNESS

Audiotapes

IDEA Seniors Specialty Course
Infomedix
12800 Garden Grove Blvd., Suite F
Garden Grove, CA 92643
(800)367-9286
In CA (800)992-9286

Max—Mature Aerobic Exercise
YMCA
340 Montgomery Street
Syracuse, NY 13202

Books

Aging in Action: A Dynamic Approach to Exercise
by Ann Marshall
Marshall Dynamics
P.O. Box 29605
Atlanta, GA 30359

Instruction Manual
by Teresa Young, MS
Young Enterprises, Inc.
107 N. Main
Lansing, KS 66043

Physical Activity, Aging, and Sports, Vol. I:
Safe Therapeutic Exercise for the Frail Elderly
by Olga Hurley
Center for the Study of Aging
706 Madison Avenue
Albany, NY 12208
(518)465-6927

Seniors on the Move
by Renate Rikkers
Human Kinetics Publishers
P.O. Box 5076
Champaigne, IL 61820-9971
(800)342-5457

The Forty Plus Market: How to Capture and Keep Full-Life Members
IRSA
132 Brookline Avenue
Boston, MA 02215
(617)236-1500

The Fun of Fitness: Handbook for the Senior Class
by Betty Perkins-Carpenter
Senior Fitness Productions
1606 Penfield Road
Rochester, NY 14625-0413
(716)586-7980

Organizations

Administration on Aging
330 Independence Avenue
Washington, DC 20201
(202)472-7257

American Association of Retired Persons
1909 K Street N.W.
Washington, DC 20049
(202)872-4700

Arthritis Foundation
P.O. Box 19000
Atlanta, GA 30326

National Association for Area Agencies on Aging
600 Maryland Avenue S.W.
West Wing, Suite 208
Washington, DC 20024
(202)484-7520

Products

Squeeze Ball and Fitness Fun (poster and ball)

Exer Fun
3089-C Clairmont Drive, #130
San Diego, CA 92117
(619)268-0684

Video Tapes

Exercises for Seniors
Lincoln Medical Educational Foundation
4600 Valley Road
Lincoln, NE 68510
(402)483-4581

Exercise Seniorstyle
by Susan Malmsadt Wanner and Marilynn Freier
Sunrise Media
96 Inverness Drive E
Englewood, CO 80112
(800)827-3726

Fitness Over 50
General Television Network
Health Tapes, Inc.
13225 Capital Avenue
Oak Park, MI 48237
(313)548-2500

Focus on Fitness
with Tena Eddy-Pulley
Downtown Atlanta Senior Services
607 Peachtree Street, NE
Atlanta, GA 30365
(404)872-9191

More Alive
by Jo Murphy
Mature Adult Corporation
P.O. Box 98
Lafayette, CO 80026
(800)873-3347

Older Adult Exercise Program
Young at Heart Life Time Fitness
107 N Main
Lansing, KS 66843
(913)727-2263

Senior Shape Up
Yablon Enterprises, Inc.
P.O. Box 7475
Steelton, PA 17113-0475
(713)939-4545

Seniorcise
by Terry Ferebee
Trinity Wellness Center
P.O. Box 5020
Minot, ND 58701

Silver Foxes II
JCI Video
5308 Derry Drive
Agoura Hills, CA 91301
(818)889-9022

Swing into Shape
Lutheran Hospital
1910 South Avenue
La Crosse, WI 54601
(608)785-0530 Ext. 3194

References

National Survey of Worksite Health Promotion Activities, U.S. Department of Health & Human Services, U.S. Gov. Printing Office, Washington, DC, 1987.

Promoting Health/Preventing Disease: Year 2000 Objectives for the Nation, Public Health Services, U.S. Department of Health and Human Services, U.S. Gov. Printing Office, Washington, DC, 1989.

Health Promotion Cost Versus Benefits, The University of Michigan Fitness Research Center, Ann Arbor, MI, 1990.

Modern Nutrition in Health and Disease, Goodhart, 1978.

Healthy Worker—Healthy Workplace, Senate Committee on Industrial Relations, California Legislature, 1990.

Loss Control for the Small to Medium Size Business, Robert E. Brisbin, Van Nostrand Reinhold, New York, 1990.

Best's Underwriting Guides, A.M. Best Company, 1990.

Industrial Hygiene: A Training Manual, Richard S. Brief, Exxon Corp. Medical Dept., Linden, NJ, 1975.

Housekeeping Handbook For Institutions, Business and Industry, Edward B. Feldman, Frederick Fell Inc., New York, 1969.

Industrial Safety Handbook, William Handley, McGraw-Hill, New York, 1970.

Supervisors Guide to Human Relations, Earle S. Hannaford, National Safety Council, Chicago, IL. 1976.

Supervisory Training and Development, Donald L. Kirkpatrick, Addison-Wesley Publishing Co., Reading, MA, 1971.

Motivation and Personality, Abraham H. Maslow, Harper and Row, New York, 1970.

Index

Achievement logs, 108
Activity programs, in wellness program, 22
Aerobic dance, 41-42
Aerobic fitness facility, 43-44
    equipment in, 43
Alcohol and controlled substance rehabilitation, x, 51-54
    consultative services, providing education on, 15
    and controlled substance testing in preemployment physical examination, 53, 77-78
    exercise in, 33
    federal guidelines in, 52
    importance of legal advice in, 52
    as performance issue, 52
    referrals of candidates testing positive to, 78
    and zero tolerance substance abuse policy, 51-53, 78
American College of Obstetricians and Gynecologists, guidelines for exercise during/following pregnancy, 68-69
American College of Sports Medicine (ACSM), 7, 11
Amino acids, 135-138
    supplements, 137-138
Appointment notification, for fitness profile, 113
Arm rests, for chair, 97
Association for Fitness in Business (AFB), 7

Back care, 101-105
    consultative services for, 14-15
    muscle strain injuries in, 102-105
    spinal injuries in, 101-102
    weight lifter belts in, 103-105
Back injury, use of employee health benefit systems for off-site, 93
Back support belts, 103-105
Biofeedback training, in stress management, 64
Blood sugar levels, 59
Body composition, monitoring, 60
Budget, for wellness program, 25-30

Cafeteria
    food choices in on-site, 58-59
    management of, in improving nutrition, 59
Calisthenics. *See also* Exercise; Fitness program
    and avoiding strain injuries, 103
Captive insurance company, 82-83
Carbohydrates, 138-139
Cellulose, 139
Chair
    adjustable height for, 97, 100
    ergonomic factors in selecting, 96-98
Child care, 70-71
Child care center
    employment of senior citizens in, 70-71
    establishment of, at workplace, 70
    negotiations of discount at, 71
Circuit training, 42-43
Clothing, appropriate, and back care, 103

150 Index

Communications analysis, in stress management, 64
Community, availability of wellness program to, 29
Community center, for off-site fitness program, 44
Community fun run events, employer sponsored, 39
Community services, integration into wellness program, 28-29
Compensable trauma, definition of, 92-93
Computer workstation ergonomics, 93, 94-96, 99-100
  furniture in, 96-99
Confidential information, verification of, in hiring, 73-76
Consultant
  choosing, as wellness director, 8
  qualifications of outside, 8-10
Consultative services, specialized, 10-15
Controlled substances testing, in preemployment physical examination, 53, 77-78
Cumulative trauma
  associated with computer workstations, 94
  growth of claims for, 92

Daily activities, incorporation of stress management into, 62-63
Dainoff, Marvin, 94
Designated physician program, 78
Diet. *See also* Nutrition; Weight management and nutrition education
  integration into wellness program, ix-x
Dietary approaches, problems with artificial, 57-58
Dietician, use of outside, in improving nutrition, 59
Direct writing, as source of insurance quotes, 87-88
Document holders, 98, 100
Drug and alcohol education and rehabilitation. *See* Alcohol and controlled substance rehabilitation
Drug screening
  in pre-employment physical examination, 53, 77-78
  sample policy in, 52-53
  in work related accident investigations, 53

Early return to work programs, and workers' compensation, 79-81
Elder care, 71-72. *See also* Senior citizens
  resource qualification and assistance, 71-72
Employee drug screening policy. *See also* Drug screening
  sample, 52-53
Employee funded modules, 29
  availability to community, 29
  stipulated fee for, 29
Employee health coverage plans, 24 hour, 27-28
Employee health screenings, 25-26
Employee incentives and recognition, 106-109
Employee interest surveys, 21
  sample, 117-118
Employee selection, 73-76
Employees, training, as wellness module instructors, 29
Environmental audit, sample, 123
Environmental Protection Agency, designation of second-hand smoke as class A carcinogen, 48
Equal Employment Opportunity Commission (E.E.O.C.)
  and hiring policies, 75
  need for guidelines from, on drug screening, 53
Ergonomics, 92-105
  and back care, 101-105
  and computer workstations, 93, 94-100
  definition of, 92
  and job analysis, 95
  and loss control, 93
  and posture, 95-96, 99-100
Errors and omissions insurance, 91
Evaluation
  of running programs, 40
  of smoking cessation programs, 50-51
  of walking programs, 38-39
  of weight management programs, 59-60
  of wellness programs, 22
Exclusive agent, as source of insurance quotes, 87
Executive oversight committee, in hiring policy, 76
Exercise. *See also* Fitness program
  health benefits of regular, ix

Index   151

participation in, and informed consent, 110–111
in prenatal wellness programs, 68–69
in smoking cessation programs, 32–33
in stress management, 32
in substance abuse programs, 33
in weight management, 59, 60–61
in wellness programs, ix

Fats, 139–140
Fat soluble vitamins, 140
Female employees, fitness skill classes for, 34
Fiber, 139
Fitness certification, organizations offering, 11
Fitness instructor
 need for qualified, 36–37
 professional affiliations for, 12
 qualifications of, 11–12
Fitness profile
 appointment notification of, 113
 and informed consent, 111–113
Fitness program, 31–47. *See also* Exercise
 consultative services providing, 11–12
 goal setting in, 33
 health and fitness attitude on participants, 31–32
 individualization in, 31
 off-site options, 34, 44–45
 on-site options, 34–44
  aerobic dance, 41–42
  aerobic fitness facility, 43–44
  circuit training, 42–43
  exercise leaders, 36–37
  fitness skill classes, 34–36
  running programs, 39–40
  stretching, 40–41
  walking programs, 37–39
 recreational programs, 45–47
 safety conscious attitude in, 31
 skill classes, 34–36, 38, 39, 42
 stress management in, 64
 variety in, 33
Fitness training. *See* Weight training
Fluorescent lighting, 98
Footrests, 95–96, 100
Forearm rests, 95, 97
Free weights, 132–133
Frequency rates, 126–129

Furniture, ergonomic factors in selecting, 96–99

Goal setting
 in fitness program, 33
 in wellness program, 22–23
Groups, for stress management, 63, 65–66

Health and fitness assessment, 21
Health benefits utilization, use of frequency rate formula in tracking, 127
Health care costs
 for illness, 1–5
 increased, due to smoking, 50
 methods of tracking, 125–129
 reducing, 1–5
Health clubs
 equipment and activities at, 45
 off-site fitness program at private, 44
 wellness program offered through, 16
Health/fitness assessment, for participants in weight management program, 60
Health risk appraisal and lifestyle profiles
 family history, 115–116
 lifestyle, 116–117
 personal history, 114–115
 resources for appraisals, 142–143
Hiring, 73–76
 consistency of policy in, 73–74
 and preemployment physical examinations, 76–78
 verification of confidential information, 73–76

Illness, cost of, 1–5
Independent insurance agency, 85–86
Independent insurance agent, 85–87
Independent insurance broker, 85–87
Individualization, importance of, in fitness program, 31
Infinite resistance machines, 133
Informed consent
 and exercise participation, 110–111
 and fitness profile, 111–113
Instructor guided support groups, 57
Instructors
 for aerobic dance, 41
 for circuit training, 43
 for community programs, 44
Insurance companies, wellness program offered through, 16

Insurance options, 82–91
  choosing agencies in, 87–88
  cost reductions in, 83–85
  customizing insurance contracts in, 89–90
  direct writing system, 87–88
  exclusive agent system, 87
  independent insurance agency as, 85–87
  lead time in obtaining quotes, 87
  self-insured option, 90–91
  types of, 82–83
Insurance policies, team participation, 47
Interest questionnaires, 25
  in determining food preferences, 59
  sample, 119–120, 122–123
International Dance Exercise Association, 11

Job description, correlation of preemployment physical examination with, 76–77

Keyboard rests and mounts, adjustable, 99

Lifestyle factors, viii–x. *See also specific factors.*
Lighting, 98–99
Lipids, 140
Loss control
  and ergonomics, 93
  program to reduce workers' compensation claims, 84
Loss exposure, and placement of insurance, 85–86
Low frequency radiation, health risks of, 99
L-tryptophan, 137
Lumbar support, in chair, 96

Male employees, fitness skill classes for, 34–35
Malpractice, 27
Management
  sample, survey of, 123–124
  support for stress management program, 64–65
  support for wellness program, 18, 19–20
Manual rate, 84
Massage therapy, in stress management, 64
Medical facilities, wellness program offered through, 16
Modifiable risk factors, viii. *See also specific factors.*
Muscle strain, 102–105

National Strength and Conditioning Association, 11
Natural lighting, 98–99
Neck strain, reducing, 98
Needs assessment, 25–26
  determining preferences for on-site versus off-site fitness programs, 45
  in wellness program, 20–22
Nutrition. *See also* Weight management and nutrition education
  amino acids, 135–138
  carbohydrates, 138–139
  class in, for workplace population, 55–56
  fats, 139–140
  fiber, 139
  information resources on, 143–144
  integration of stress management into program for, 64
  in prenatal wellness program, 68–69
  trace minerals, 140–141
  vitamins, 140–141

Occupational and industrial medicine clinics, x–xi
Off-shore captive chartering, 82–83
Off-site fitness programs, 44–45
Olympics style competition, 47
One card system, 130–131
On-site cafeteria, food choices in, 58–59
On-site fitness programs, 34–44
  aerobic dance, 41–42
  aerobic fitness facility, 43–44
  circuit training, 42–43
  exercise leaders, 36–37
  fitness skill classes, 34–36
  running programs, 39–40
  stretching, 40–41
  walking programs, 37–39

Pectin, 139
Peer pressure, in controlling substance abuse, 54
Perinatal care, resources on, 144–145
Personality based programs, in stress management, 64
Physician
  choice of, in work related illness or injury, 78–79
  need for evaluation, in exercise program, ix

Index

perspective of, vii–xi
traditional role of, in workplace, x–xi
Positive reinforcement training, in stress management, 64
Posture, at computer workstations, 95–96, 99–100
Preemployment physical examinations, 73, 76–78
  controlled substances testing in, 53, 77–78
  correlation with job description, 76–77
Pregnant employee
  prenatal wellness program for, 67–70
  resources for, 144–145
Prenatal care, and informed consent, 110
Prenatal wellness, 67–70
  exercise and nutrition in, 68–69
Prescription drugs, impact of, on controlled substances testing, 77
Private health club
  equipment and activities at, 45
  for off-site fitness program, 44
Proactive workplace, prevention of injuries/illnesses in, viii
Productivity, impact of child care worries on, 70
Product liability insurance, 91
Professional affiliations
  for wellness consultant, 10
  for wellness director, 6–7
Professionals, use of outside, in stress management programs, 63, 64

Radiation, health risks of low frequency, 99
Reactive system, disadvantages of, vii–viii
Recognition, as key word in incentive programming, 107–108
Recreational programs, 45–47
Relaxation techniques, in stress management, 64
"Rent a captive," 83
Risk management assessment, 90–91
Running programs, 39–40
  evaluation of, 40

Safety conscious attitude, importance of, 31
Safety tips, for weight training, 134–135
Screening evaluation, sample, 119
Second-hand smoke, designation of, as class A carcinogen, 48
Self-esteem, and weight management, 56–57

Self-insured insurance option, 82, 90–91
Senior citizens. *See also* Elder care
  employment of, in child care facilities, 70–71
  resources on fitness of, 146–147
Shoulder strain, reducing, 98
Single service providers, evaluating, 10–11
Skill building classes, 34–36
  for aerobic dance, 42
  content of, 35–36
  for running/jogging program, 39
  for walking program, 38
Smokefree workplace policy, 49
  development and implementation of, 48–49
Smoking, risks of, viii–ix
Smoking cessation programs, 48–51
  consultative services providing, 15
  convenience of participation in, 50
  critical periods in, 51
  evaluation of, 50–51
  and exercise, 32–33
  follow-up on, 51
  goal in, 48
  off-site versus on-site, 50
  resources on, 144
  standalone, 50
  voluntary participation in, 49–50
Specialized consultative services, 10–15
  in back care, 14–15
  in drug and alcohol education and rehabilitation, 15
  in fitness, 11–12
  in nutrition education and weight management, 12–13
  in smoking cessation, 15
  in stress management, 13–14
Spinal column claims, 92
Spinal disks, 101–102
Standalone stretching class, 40
Strain injuries
  avoiding, and calisthenics, 103
  cumulative nature of, 103
Stress
  and child care worries, 70
  definition of, 62
  and elder care, 71
Stress evaluation survey, 65
Stress management, x, 62–66
  consultative services providing, 13–14

Stress management (*cont.*)
  in daily activities, 62–63
  and exercise, 32
  goal of program, 62
  group formation, 65–66
  information resources on, 144
  management support for, 64–65
  in weight management/nutrition education program, 59
  in wellness program, 64
Stressors, cumulative effect of, 63
Stress perception evaluations, 63–64
Stretching exercises, 40–41
  and ergonomics, 95
  work specific, 41
Subcontractors, use of, by consulting firm, 9
Substance abuse, criminal aspects of, 54
Substance abuse programs. *See* Alcohol and controlled substance rehabilitation
Support groups, in weight management, 57, 61
Survey questions, 110–124

Team participation insurance policies, 47
Team sports
  liabilities associated with, 46
  opportunities offered by, 46
  recognition awards in, 107
  scheduling of, 46
Tobacco education program. *See* Smoking cessation programs
Trace minerals, 140–141
Training program, in back care, 101
Treatment follow-up, and workers' compensation, 79

Variable resistance stacked weight machines (VRSW), 133
Variable stacked weight machines (VSW), 133
Variance charts, 129–130
Vending machine content, analysis and evaluation of, 58
Video display glare screens, 98–99
Vitamins, 140–141

Walking programs, 37–39
  evaluation of, 38–39
  recognition awards for, 108
  and weight management, 61
Water immersion measurement, 60
Water soluble vitamins, 140
Weight and fitness training equipment, purchase of, 26
Weight lifter belts, 103–105, 135
Weight management and nutrition education, 55–61
  consultative services providing, 12–13
  emphasis of overall health factors in, 56
  focus of, 56
  and interrelated exercise program, 59
  monitoring success of, 59–60
  offering of nutrition classes in, 55–56
  and stress management, 59
  successful long-term, 55
  and walking program, 38
Weight training, 132, 134
  free weights in, 132–133
  infinite resistance machines in, 133
  safety tips for, 134–135
  variable resistance stacked weight machines in, 133
  variable stacked weight machines in, 133
Wellness budget, 25–30
Wellness Councils of America (WELCOA), 7
Wellness director
  choosing outside consultant as, 8–10
  commitment of, 7
  consultant guideline summary, 16
  finding, 7–8
  lifestyle of, 7
  professional affiliations of, 6–7
  qualifications of, 6–7
  role of, in monitoring off-site program, 44–45
  role of, in physician's practice, xi
  and specialized consultative services, 10–15
  work experience of, 6
Wellness module instructors, employees as, 30
Wellness policy
  developing, 17–24
  sample, 17–18
Wellness program
  activity programs in, 22
  areas addressed by, 4
  choosing right, 15–16

components of, 23-24
correlation chart, 130
employee funded modules in, 29
employee health screenings in, 25-26
employee objections to, 107
employee participation in, 29, 30, 106
evaluation of, 22
as form of loss control, 28
goal setting in, 22-23, 56
integration of community services into, 28-29
interest survey in, 25
management support for, 18, 19-20
modifiable risk factors relevant to, viii-x
motivational aspects of, 106-109
need assessment in, 20-22, 25-26
phases of, 20-22
in reducing workers' compensation claims, 84-85
tracking results in, 26-27
worker participation in, 18-20
Wellness survey, sample, 121-122
Wellness trends, variance charts in tracking, 129-130
Work breaks, 100
Workers
impact of exercise on productivity of, ix
participation in wellness program, 18-20
Workers' compensation claims
changes in injuries covered, 92
correlation of workplace wellness programs with, 85
for cumulative exposure to second-hand smoke, 48-49
and early return to work programs, 79-81
growth of cumulative trauma claims in, 94
loss control program in reducing, 84
monitoring, 78-79
for muscle injuries, 92
for spinal column claims, 92
team sport related injuries in, 46-47
and treatment follow-up, 79
warmup and stretching activities in reducing, 41
workplace wellness program in reducing, 84-85
Workers' Compensation Insurance, premiums paid for, 84
Work force, characteristics of healthy, 4
Workplace environmental factors, government regulation of, viii
Workplace medical care, common scenario of, xii
Work related accident investigations, drug screening in, 53
Work specific stretching, 41
Work stations, adjustable, 100
Work surface lighting, supplemental, 98
Work surfaces
adjustable, 99
height of, 97
Wrist disorders, 94
Written work release, in returning to work, 79-80

Yoga, 40-41

Zero tolerance substance abuse policy, 51-53, 78